in **DETAIL** Cost-Effective Building

in **DETAIL**

Cost-Effective Building

Economic concepts
and constructions

Christian Schittich (Ed.)

Edition DETAIL – Institut für internationale
Architektur-Dokumentation GmbH
München

Birkhäuser
Basel · Boston · Berlin

Editor: Christian Schittich
Editorial services: Astrid Donnert, Alexander Felix, Carola Laule
Michaela Linder, Melanie Schmid, Cosima Strobl, Edith Walter

Translation German/English:
Catherine Anderle-Neill

Drawings: Nicola Kollmann, Marion Griese, Daniel Hajduk,
Caroline Hörger

DTP: Peter Gensmantel, Andrea Linke, Roswitha Siegler, Simone Soesters

A specialist publication from Redaktion DETAIL
This book is a cooperation between
DETAIL – Review of Architecture and
Birkhäuser – Publishers for Architecture

Library of Congress Control Number: 2007927593

Bibliographic information published by the German National Library
The German National Library lists this publication in the Deutsche
Nationalbibliografie; detailed bibliographic data is available on the Internet at
<http://dnb.d-nb.de>.

This book is also available in a German language edition
(ISBN: 978-3-7643-8393-0).

© 2007 Institut für internationale Architektur-Dokumentation GmbH & Co. KG,
P. O. Box 33 06 60, D-80066 Munich, Germany and
Birkhäuser Verlag AG, Basel · Boston · Berlin, P. O. Box 133, CH-4010 Basel,
Switzerland

This work is subject to copyright. All rights are reserved, whether the whole or
part of the material is concerned, specifically the rights of translation, reprinting,
re-use of illustrations, recitation, broadcasting, reproduction on microfilms or in
other ways, and storage in data banks. For any kind of use, permission of the
copyright owner must be obtained.

Printed on acid-free paper produced from chlorine-free pulp (TCF ∞)

Printed in Germany
Reproduction:
Martin Härtl OHG, München
Printing and binding:
Kösel GmbH & Co. KG, Altusried-Krugzell

ISBN: 978-3-7643-8393-0

9 8 7 6 5 4 3 2 1

Contents

Cost and quality awareness in building – a challenge Christian Schittch	9
Project Summary	10
Showroom near Zevenbergen ott architekten, Augsburg	12
Cost efficient but not cheap Wolfgang Ott	16
Studio in Madrid Abalos & Herreros, Madrid	24
Pharmacy and Medical Practice in Plancher-Bas Rachel Amiot & Vincent Lombard, Besancon	28
Static Lightness in Timber Jean Luc Sandoz	32
Mobile House in England Mae Architects, London	34
Prefabricated House from Denmark ONV arkitekter, Vanløse	38
Straw House in Eschenz Felix Jerusalem, Zurich	42
House in Shimane Sambuichi Architects, Hiroshima	46
House in Aitrach SoHo Architektur, Augsburg	50
Weekend House in St Andrews Beach Sean Godsell Architects, Melbourne	54
House on Lake Laka Peter Kuczia, Osnabruck	58
Housing Development in Neu-Ulm G.A.S.-Sahner, Stuttgart	62
Different Forms of Construction for a Modular Unit Building System Georg Sahner	66
Terrace Housing in Milton Keynes Rogers Stirk Harbour + Partners, London	70
Apartment House in Dortmund ArchiFactory.de, Bochum	72
Housing in London Ash Sakula Architects, London	76
Multi-Storey Housing in London Niall McLaughlin Architects, London	78
Multi-Storey Housing in Munich Hierl Architekten, Munich	84
Cost and Quality Awareness in Construction Rudolf Hierl	88
Hall of Residence in Amsterdam Claus en Kaan Architecten, Amsterdam	92
Youth Camp in Passail Holzbox Tyrol, Innsbruck	96
Hotel in Groningen Foreign Office Architects, London	102
Cultural Centre in Munich Ingrid Amann Architekten, Munich	106
Cost efficient building means sustainable building Ingrid Amann	112
Secondary School in Brixlegg Raimund Rainer, Innsbruck	114
Secondary School in Eching Diezinger & Kramer, Eichstätt	118
Montessori College Oost in Amsterdam Herman Hertzberger, Amsterdam	122
Mini-golf and Ice Skating Hall in Bergheim mfgarchitekten, Graz	128
Ice Stadium in Wolfsburg Schultz + Partner, Braunschweig	130
"Economy does not mean building cheaply" Helmut C. Schultz	134
Service Centre in Frankfurt a. M. Dietz Joppien, Frankfurt a. M.	136
Planning as „refined carcass" construction: economical, functionally neutral, robust Matthias Schönau	140
Office and Training Centre in Dresden Heinle, Wischer and Partner, Dresden	142
Production Building in Großhöflein querkraft architekten, Vienna	146
Cost Management querkraft architekten with Erwin Stättner	151
Factory extension in Murcia Clavel Arquitectos, Madrid	152
Wiper Factory in Bietigheim-Bissingen Ackermann and Partner Architekten, Munich	156
From Urban Planning to Construction Grid Peter Ackermann	160
The Optimisation of the Load-Bearing Structure Christoph Ackermann	162
The Building in Operation Helmut Bucher	164
Architects – Project details	166
Illustration Credits	176

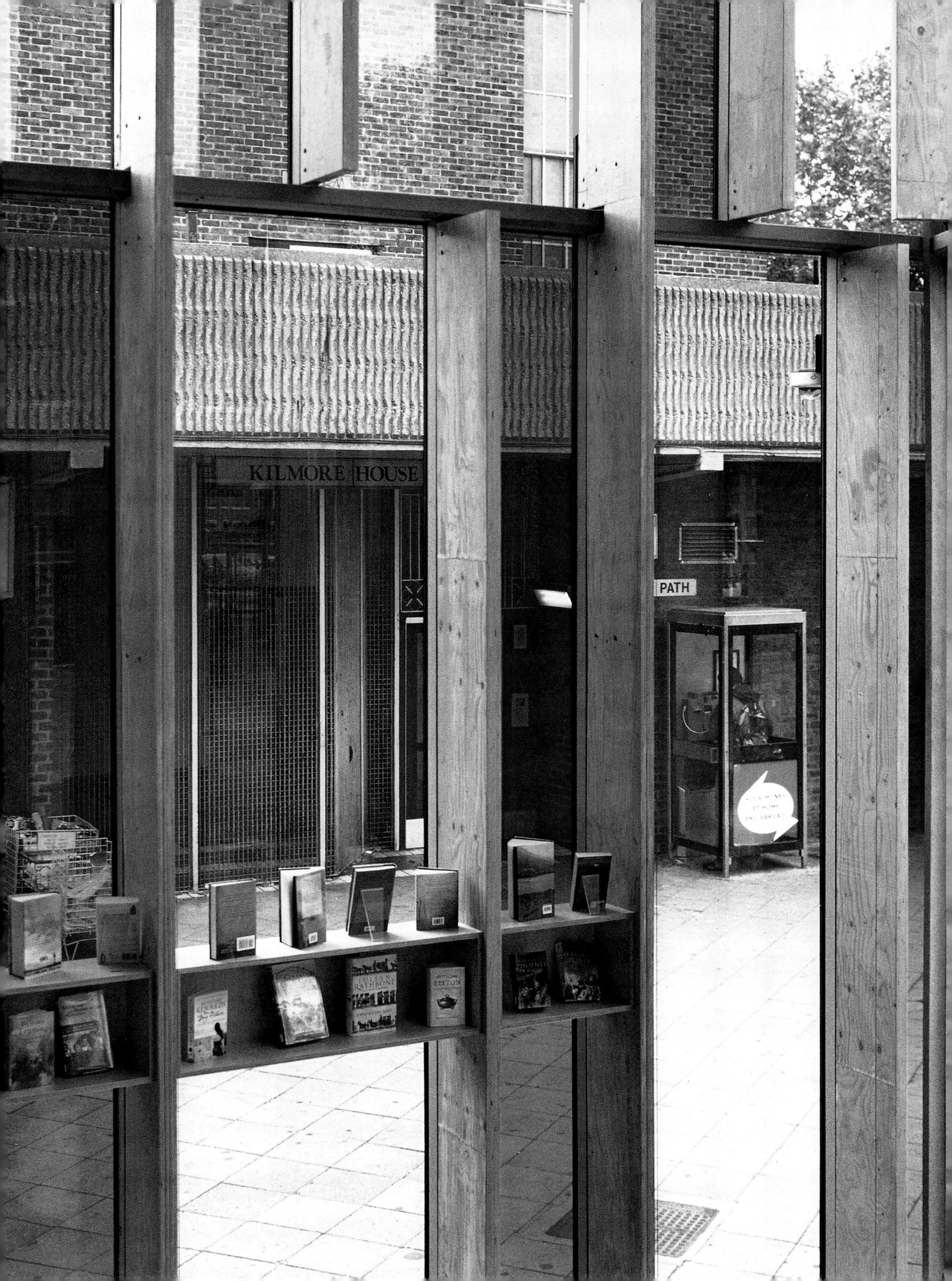
KILMORE HOUSE
PATH

Cost and quality awareness in building – a challenge

Christian Schittich

Spectacular constructions created with often exorbitant budgets are the popular image of architecture as seen in the media today. When cities or corporations, particularly those from the automobile industry, are presenting themselves to the world, the question of funding seems to be of no consequence. Famous architectural stars are provided with apparently endless resources in order to attract attention by creating simply stunning objects. For the majority of architects and planners, however, the day to day reality is something else. Costs and budget limitations are the rule, generous resources being a rare luxury. Designers are continuously presented with new challenges in the on-going struggle for the best, most economical and pleasing solution for the lowest possible price.

Cost efficiency is not the same as cheap building; but it must not by definition be a disadvantage. Often doing away with the multitude of superfluous elements can lead to a more aesthetically credible solution. For the planner, however, cost efficient building usually means an increase in concerted effort which is seldom, if ever, adequately remunerated. Efficient office structure and self-discipline are of paramount importance. Generally speaking, though, the increased economic awareness creates a new opportunity for the profession. Namely, the establishment of the architect as all-encompassing expert responsible for economical solutions to everything from initial concept through construction, to subsequent management and energy consumption control.

Cost efficient building sometimes even starts with the initial selection of location affecting, as it does, the type of development, the construction site and subsequent energy consumption as a result of prevailing wind directions and solar gains. At the very latest, however, cost efficiency begins in the preliminary planning stage of the project; with the concept. The optimisation of functional sequences and written brief, the built form, the relationships between volumes and facade areas, and whether a basement is necessary are just some of the relevant parameters. Finally, in the construction stage, cost efficiency depends to a great extent on the selection of materials, the most economical structural system and detailing, and the question of whether prefabricated elements are feasible or not.

With this book and the buildings within, we intend to direct attention to a particular aspect of architecture which, though neither overly elaborate nor fashionably austere, demonstrates exactly where the essence of well designed yet economical building solutions can be found. These constructions comply with the relevant local standards, although simply fulfilling requirements can not be considered a determining factor in the selection process. Much more than that, a publication which presents and compares international building examples should also make the effort to query one or another regulation or edict which tends to force costs up. Data pertaining to construction costs must necessarily be included where economy is being considered. Wherever possible, these have been presented with the relevant projects, although it must be realised that it is often difficult to compare them. Place and date of construction, in addition to various other factors all bear relevance; the price of a square metre in London which is remarkably economical would appear exorbitant in Berlin. However, cost information offers valuable reference points, sometimes assisting in the comparison of different construction processes within a single project. Because cost efficient building can really only be demonstrated by means of practical examples, some of the architects and consultants responsible for the examples presented in this book have also contributed articles. Concise descriptions of the individual projects are presented here, in addition to longer discussions of personal strategies, whether relating to particular examples or to cost efficiency in general.

Project Summary

Page	Project	Areas/Volumes	Cost of Construction	Construction	Detailed Breakdown of Costs
12	Showroom near Zevenbergen ott architekten, Augsburg	406 m² TFA 380 m² UA 1,624 m³ TIV	450,000 € 1,108 €/m² TFA	steel frame + stacked plank elements	construction with general contracto no cost breakdown available
24	Studio in Madrid Ábalos & Herreros, Madrid	302 m² TFA 328 m² UA	500,000 € (total) 1,524 €/m² UA	reinforced concrete + steel	
28	Pharmacy and Medical Practice in Plancher-Bas Rachel Amiot & Vincent Lombard, Besançon	615 m² TFA 560 m² UA 1,940 m³ TIV	670,000 € (net) 800,000 € (total) 1,196 €/m² UA 115 €/m² roof area	timber	excavation/foundation 4.1% external/internal walls 32.2% slabs/roof 27.5% services 10.1% technical facilities 26.1%
34	Mobile House in England Mae Architects, London	86 m² TFA 215 m³ TIV	195,000 € (total) 1,950 €/m² UA	timber or steel framework	
38	Prefabricated House from Denmark ONV arkitekter, Vanløse	60–169 m² TFA	1,400 €/m² UA	timber framework	
42	Straw House in Eschenz Felix Jerusalem, Zurich	173 m² TFA 140 m² UA 844 m³ TIV	426,738 € (total) 2,463 €/m² UA	timber	
46	House in Shimane Sambuichi Architects, Hiroshima	271 m² UA 722 m³ TIV		timber	
50	House in Aitrach SoHo Architektur, Augsburg	172 m² TFA 132 m² RA 775 m³ TIVI	230,000 € (total) 1,337 €/m² TFA	timber framework	
54	Weekend House in St Andrews Beach Sean Godsell Architects, Melbourne	280 m² TFA 800 m³ TIV		steel skeleton	
58	House on Lake Laka Peter Kuczia, Osnabruck	226 m² TFA 175 m² UA 730 m³ TIV	95,500 € 422 €/m² TFA	masonry + timber	excavation/foundation 2.5% external/internal walls 35% slabs/roof 25% services 22% technical facilities 15.5%
62	Housing Development in Neu-Ulm G.A.S.-Sahner, Stuttgart	2,731 m² TFA 10,487 m³ TIV	1.82 mill. € (total) 817–941 €/m² RA	masonry	construction 82.2% TS 17.8%
70	Terrace Housing in Milton Keynes Rogers Stirk Harbour + Partners, London		1.92 mill. € (total) 1,267 €/m² UA	timber framework	
72	Apartment House in Dortmund ArchiFactory.de, Bochum	556 m² TFA 1,440 m³ TIV	200,000 € (net) 360 €/m² TFA	masonry	excavation 5% renovation 75% extension 20%
76	Housing in London Ash Sakula Architects, London	338 m² TFA 268 m² UA 918 m³ TIV	865,830 € (total) 2,202 €/m² TFA	timber framework	
78	Multi-storey Housing in London Niall McLaughlin Architects, London	1,071 m² TFA 966 m² RA 2,870 m³ TIV	2.20 mill. € 2,200 €/m² RA	timber framework	excavation/foundation 9.8% external/internal walls 28.9% slabs/roof 8.9% structure 10.5% services 16.9% technical facilities 9.6%
84	Multi-storey Housing in Munich Hierl Architekten, Munich	10,450 m² TFA 7,595 m² RA 35,371 m³ BRI	8.72 mill. € (total) 1,149 €/m² RA	reinforced concrete	
92	Hall of Residence in Amsterdam Claus en Kaan Architecten, Amsterdam/Rotterdam	3,650 m² TFA 9,000 m³ TIV	3.6 mill. € (total) 986 €/m² TFA	reinforced concrete + exposed masonry	excavation/foundation 3.2% carcass 58.9% fit-out 9.1% heating/air 11.3% electrics 6.3% additional costs 11.2%

Page	Project	Areas/Volumes	Cost of Construction	Construction	Detailed Breakdown of Costs
96	Youth Camp in Passail Holzbox Tyrol, Innsbruck	550 m² TFA 470 m² UA 1,800 m³ TIV	708,000 € (net) 1,500 €/m² UA	timber panel system	basic module 48.4% plinth 16.9% wall layers 2.4% roof layers 3.7% furniture 15.6% services 2.8% external paving 5.3% landscaping 4.9%
102	Hotel in Groningen Foreign Office Architects, London	210 m² TFA 598 m³ TIV	450,000 € 2,142 €/m² TFA	steel	
106	Cultural Centre in Munich Ingrid Amann Architekten, Munich	1900 m² TFA 1705 m³ UA 9478 m³ TIV	1.8 mill. € 1,100 €/m² UA	reinforced concrete	
114	Secondary School in Brixlegg Raimund Rainer, Innsbruck	4,100 m² TFA 3,441 m² UA 16,245 m³ TIV	4 mill. € (net) 1,162 €/m² UA	reinforced concrete	
118	Secondary School in Eching Diezinger & Kramer, Eichstätt	12,630 m² TFA 59,580 m³ TIV	13.8 mill. € (total) 1,090 €/m² TFA	reinforced concrete	
122	Montessori College Oost in Amsterdam Herman Hertzberger, Amsterdam	17,016 m² UA	15.2 mill. € (total) 891 €/m² UA	reinforced concrete	
128	Mini-golf and Ice Skating Hall in Bergheim mfgarchitekten, Graz	855 m² UA 5,130 m³ TIV	368,000 € 430 €/m² UA	timber skeleton	
130	Ice Stadium in Wolfsburg Schulitz + Partner, Braunschweig	10,540 m² TFA 76,720 m³ TIV	7.5 mill. € (net) 683 €/m² TFA (net)	steel	construction 4.5 mill. € (net), of which: reinforced concrete 41% steel construction/roof 22% facade 14% interior fit-out 18.3% TS 2.7 mill. € (net), of which: heating/air/plumbing 54.3% electrics 37.1%
136	Service Centre in Frankfurt am Main Dietz Joppien, Frankfurt am Main	21,750 m² TFA 13,300 m² UA 88,000 m³ TIV	14.6 mill. € (net) 1,097 €/m² UA	steel + lightweight concrete	excavation/foundation 8% external/internal walls 36% slabs/roof 32% site facilities 6% heating/air/plumbing/electrics 18%
142	Office and Training Centre in Dresden Heinle, Wischer and Partner, Dresden	2,881 m² TFA 1,465 m² UA 8,556 m³ TIV	2.8 mill. € (total) 970 €/m² TFA	reinforced concrete	cost groups 300: 1.05 mill. € (net), of which: excavation/foundation 28% external/internal walls 40% slabs/roof 32%
146	Production Building in Großhöflein querkraft architekten, Vienna	2,341 m² UA 12,989 m³ TIV	1.09 mill. € (total) 466 €/m² UA	steel	
152	Factory Extension in Murcia Clavel Arquitectos, Madrid	1,650 m² TFA 12,000 m³ TIV 1,572 m² UA	500,000 € (total) 318 €/m² UA	steel	
156	Wiper Factory in Bietigheim-Bissingen Ackermann and Partner Architekten, Munich	237,975 m³ TIV	27.5 mill. € 847 €/m² 116 €/m³	steel	

TFA = Total Floor Area, TIV = Total Internal Volume, UA = Usable Area, RA = Residential Area, TS = Technical Services.

Exhibition and Training Centre near Zevenbergen

Architects: ott architekten, Augsburg

Simple and elegant showroom
Short assembly time due to prefabrication
Economical elements

The concept foresaw the development of a commercial, multi-functional building type to contain showrooms, training spaces and offices, and which should be applicable to various locations in Europe. First implemented in the Dutch city of Zevenbergen, the building also functions as a vehicle for products and as a "brand" in its own right, the recognizable identity of the box (in the sense of corporate design) stood to the fore. The single-storey strip, roughly 10 × 40 m on plan, forms a striking volume that seems to hover just above the ground.

Minimal material selection

The fully glazed longitudinal facades of the structure provide illuminated display spaces for models of awnings and shutters. In the parapet and plinth zones, the linear windows are flanked by horizontal strips clad with anthracite-coloured fibre-cement sheeting set flush with the facade. The end walls are covered almost entirely with the same grey sheeting. Access to the exhibition container is via an external ramp or precast concrete steps set in front of one of the longer elevations. Inside the building, a spacious lounge with a landscape of bright red upholstered sofas provides a reception area that leads on to the central showrooms. In contrast to the dark-grey external skin, all internal surfaces are clad with untreated, light-coloured larch panelling. Incorporated in the soffit system is a continuous narrow lighting strip around the outer walls. The various models of awnings and shutters are presented by means of pivoting light boxes placed on the structural axes of the building. Located at the ends of the building are a training area, a store and offices.

Economical construction and assembly

The box was conceived as a mixed form of construction and had a short assembly period. By specifying a stacked-plank floor and roof, it was possible to dispense with an elaborate system of secondary beams. The load-bearing structure of the timber box consists of framed steel I-beams. Sanitary cells and spatial dividers were prefabricated in a timber-framed form of construction. After an assembly period lasting only three weeks, the exhibition building in the Netherlands was handed over ready for use.

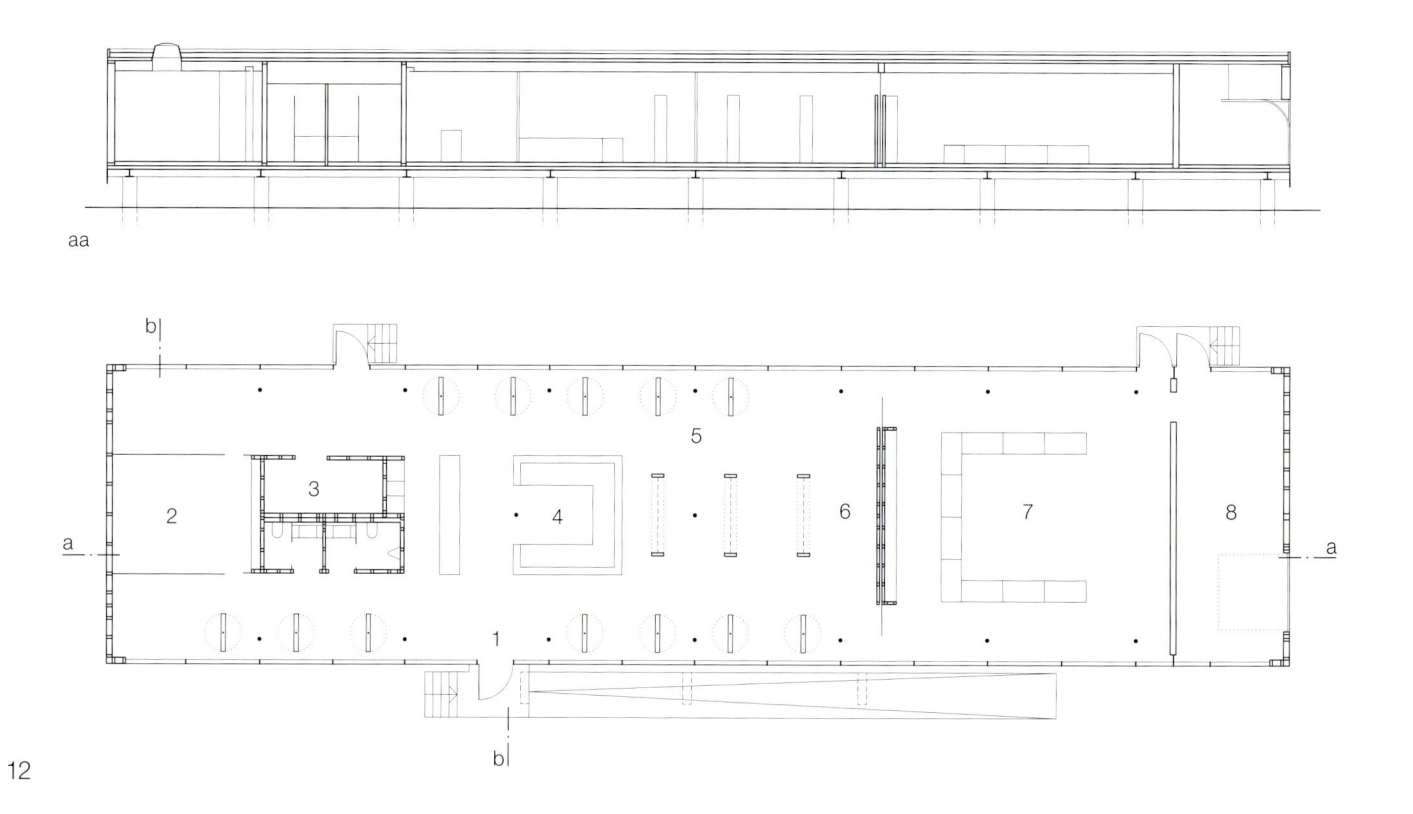

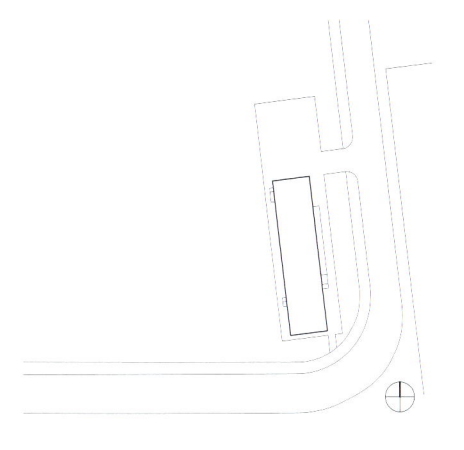

Project details:
Usage: showroom
Construction cost: 450,000 Euros
Cost per m²: 1,180 Euros
Total usable area: 379 m²
Internal ceiling height: 3.00 m
Total internal volume: 1,624 m³
Total floor area: 406 m²
Site area: 2,174 m²
Period of construction: 3 weeks
Date of construction: 2005

section • floor plan   scale 1:250
site plan   scale 1:2000

1 entrance
2 office
3 service box
4 lounge
5 exhibition area
6 spatial divider
7 training area
8 store

vertical section
horizontal section
scale 1:20

1   2 mm sheet-metal covering
2   8 mm anthracite-grey fibre-cement fascia
    40 × 50 mm battens, 2 mm sealing layer
    140 × 580 mm laminated timber beam
3   40 mm gravel, 2 mm protective mat
    2 mm roof seal, 120 mm thermal insulation
    100 mm vertically stacked-plank roof on
    300 × 320 mm laminated timber beams
    18 mm three-ply laminated-larch soffit lining,
    lime washed and brushed
4   100 mm peripheral lighting strip
5   exhibition light-box:
    20 mm black-impregnated MDF frame
    8 mm translucent white Perspex cover
6   1200 × 6000 mm balustrade element
7   180 mm precast concrete ramp slab
8   Ø 40 mm aluminium door pull
9   4 mm toughened glass bolted externally to
    50 × 120 mm anthracite aluminium RHS
    frame with 6 mm float glass
10 140 × 50 mm anthracite aluminium RHS
    window frame
11 18 mm coconut matting
12 33 mm larch floor boards, brushed and oiled

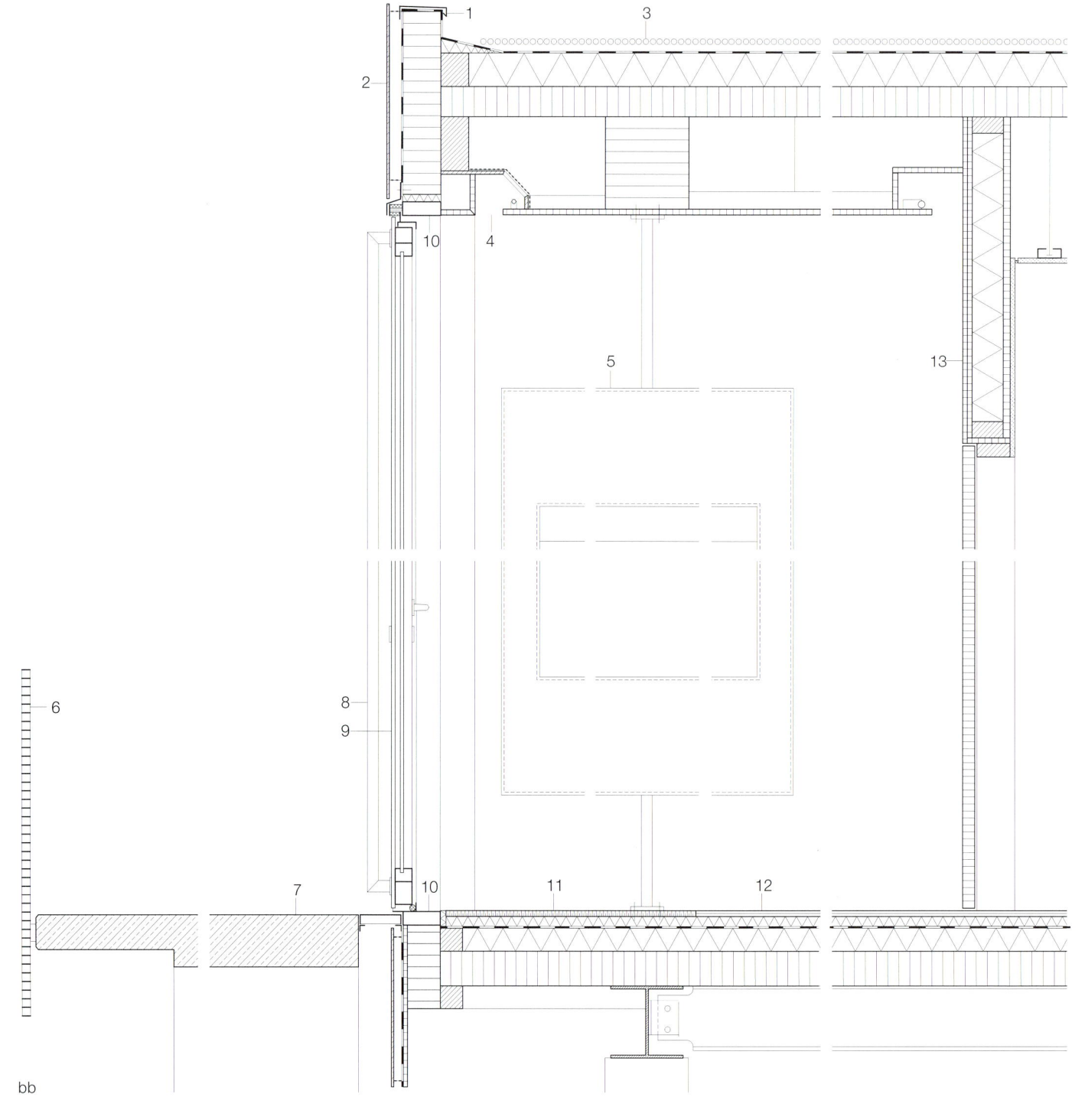

40 mm bearers,
40 mm impact-sound insulation
2 mm vapour barrier, 80 mm thermal
insulation between 80 × 80 mm bearers,
120 mm vertically stacked-plank floor
2 mm separating layer, 260 and 240 mm
steel I-beams
13 18 mm three-ply laminated larch lining
20 mm oriented-strand board, 120 mm
insulation between
120 × 60 mm wood rails, larch door
14 glazing to office unit:
10 mm toughened glass, adhesive fixed with
silicone at floor

15 Ø 70 mm tubular steel column, painted anthracite
16 4 mm toughened safety glass with
sun-protection coating +
12 mm argon-filled cavity +
6 mm toughened glass
17 8 mm grey fibre-cement sheet cladding
40 × 50 mm battens, 2 mm sealing layer
15 mm oriented-strand board
200 × 80 mm timber rails, 200 mm thermal
insulation, 2 mm sealing layer
18 mm oriented-strand board
18 mm three-ply laminated larch lining
lime washed and brushed

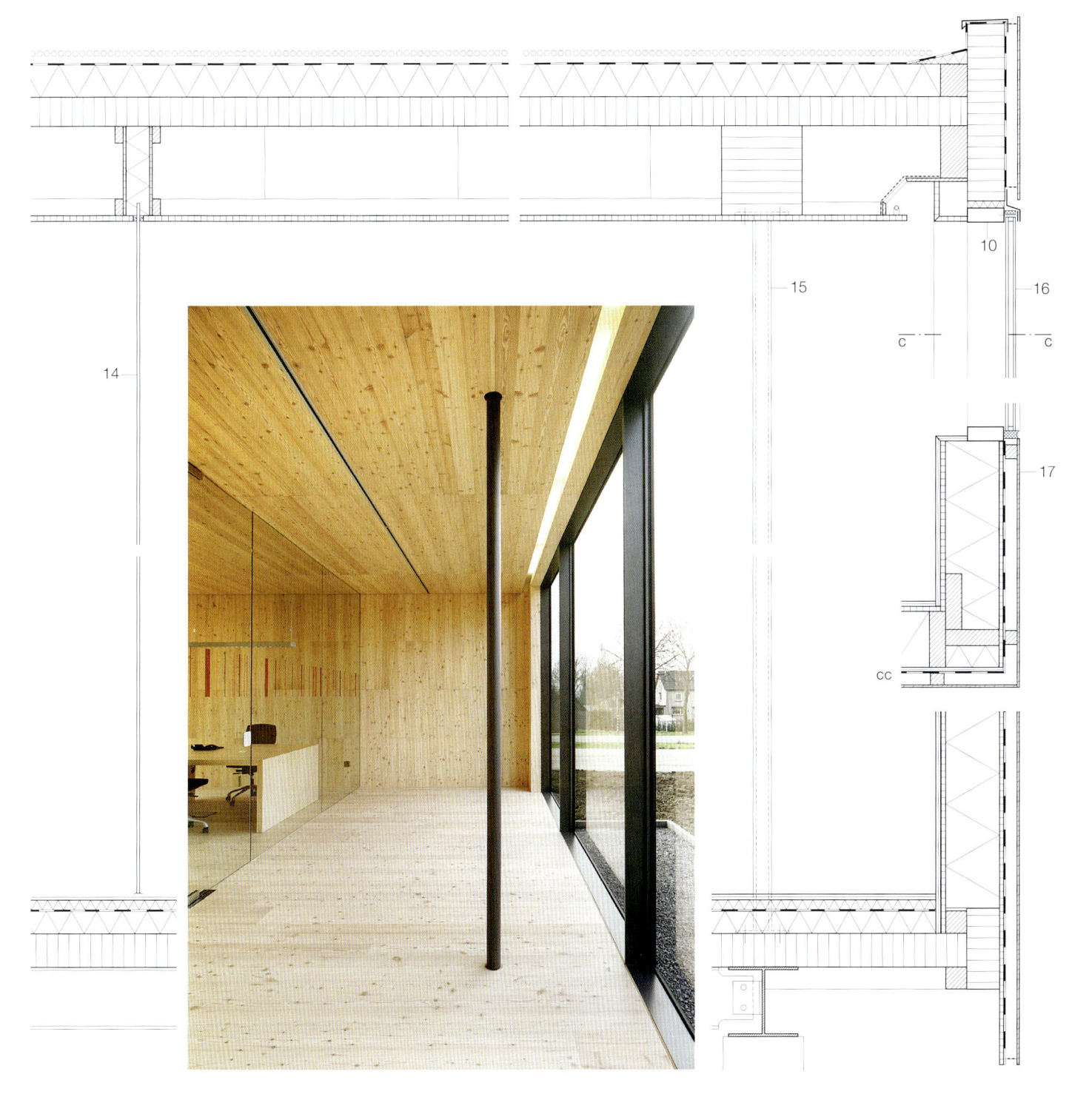

Cost efficient but not cheap

Wolfgang Ott in Augsburg

Wolfgang Ott concerns himself as much with cost efficient planning and construction as with ambitious architecture. In this interview, he discusses general concepts regard cost reduction and demonstrates them by way of projects in the Netherlands (pp. 12–15).
The activities of his office as investor have, necessarily, influence his work. Some years ago he realised the extension of a school in Nepal under his own initiative – economical construction under totally different conditions.

How, in your opinion, should the planning process be run in order to be as economical as possible?
There are always three phases for us. The first phase is quite obviously design. The design phase offers the most scope for cost saving. Every square metre of access or communication space that I don't need not only makes the architecture poorer, but also makes it notably more expensive. In the first stage, the building is honed and compacted until we are only left with what is absolutely necessary for it to function.
And, of course, the building must not only function, it must also possess architectural quality; which doesn't necessarily have to do with superfluous space, but rather to do with design aesthetics. These will be checked once again in the building application phase.

Even at the initial design stage, fire protection is of a high priority for us. We are accustomed to coordinating with fire protection agencies in order to make the design as water-tight as possible and able to be built.
The second phase contains detailed consideration of the building process. We work with prefabricated elements a lot for cost reasons; a formal effect can be also be attained with materials which are appropriate for prefabrication.
The third phase begins with the awarding of contracts. We pay close attention to the techniques of the contracted building companies rather than the product manufacturers. When planning a timber construction, for example, and the construction company which has been awarded the contract has one metre modules rather than 62.5 cm modules, it is still possible to build in a 62.5 cm grid but more time and effort will be involved. We consider the interaction with the construction companies as extremely important. It must be noted, however, that this factor is often the most difficult to take into consideration; the construction companies are selected comparatively late in the planning process and decisions must be quickly met. Reconsidering the structure between the awarding of contracts and construction begin is certainly the aspect which causes the most headaches and, of course, does not always work.

You just mentioned how important prefabrication is, but there are limitations. Where do you see them?
Prefabrication is certainly no longer efficient when a particular level of systemisation is lost. A positive element of prefabrication is the repetition, but negative aspects are the transport and assembly costs. The repetitive effect must be predominant. Prefabrication would have been inappropriate for the Waldorf School in Augsburg (ill. 3, 4), because each element was different, although we did consider it for a period of time. It is a classic in-situ concrete construction. While the "phg-Druck" building (ill. 1), here in the area, is a classic prefabricated structure, simply because it is based on elements.

But the Waldorf School was extremely economical. Does that mean that in-situ concrete constructions can be built just as economically?
No, it is not really as economical. We manages to keep the price down by convincing the client that exposed concrete doesn't have to be as smooth as possible, but can have a coarse character. If I had to build the school now with smooth formwork, it would be exorbitantly expensive and something would simply be missing. The building emanates a sturdiness which actually requires the coarseness of the concrete. Realisations like these often come later.

In which trades do you consider to hold the largest potential for cost savings?
The largest potential is certainly found in the construction of the carcass; that is the concrete, steel or timber construction, the structural engineer has great influence. We always have extra meetings when unexpected quotes for the structure come in. Then we have to sit down with the structural engineer again to find the hidden costs that we don't want to lose sight of. It starts back at the foundations again, or the locations of the columns will be questioned, perhaps a spacing of two metres is advantageous for the design but not essential. This two metre grid simply consumes a six-figure amount in the roof structure and exactly this amount is missing for the attractive finishes later. We try to balance this out; we try to control the structure to such a level that at the end of the project there is still enough money in the budget for the finishes.

Do you usually work in conjunction with external consultants?
We use external consultants. They are all offices with which we have closely worked for many years; almost like a family.

But that means that consultants can also contribute to cost savings, or are they usually responsible for higher costs?
Now I have to be a bit careful how I express myself. Consultants are not responsible for the overall budget; that is, they only have access to a particular amount of money. This budget is often met, but the total amount available for the project seldom increases over time, rather it occasionally decreases. We are usually responsible for coordinating the costs, and are always the ones who have trouble when the budget doesn't work. Our sensitivity for costs is necessarily greater than that of the consultants. Technical planners start from the comfortable situation that the building functions to a level of 150%. Structural and service engineers increase their level of risk the closer they design to the limit of functionalism. They are naturally unwilling to save in this respect, and it is our responsibility to localise this grey zone. It is here that new possibilities are created for architecture.

The way you present the processes, it sounds as if cost efficient building is exceedingly complex in the planning. Is it efficient for the architectural office?
Yes that's right, particularly if I don't come to a financial agreement with the client regarding the cost efficiency of the project. That is actually quite difficult to determine, although the German fees structure takes it into account. If I take my job seriously and consider it my responsibility to plan the building as efficiently as possible; I also make a smaller profit for my office.

So how do you balance that out in your office, with a practised team or through routine, because so many elements repeat themselves?
Over the last ten years we have built many retail and other types of constructions, and it has been necessary to develop routines in order not to lose track of the architectural quality for which we are known. On the other hand we have a relatively high proportion of clients who are extremely interested in cost efficiency. We have successfully maintained this balancing act for the last ten years and, as a result, have an extensive portfolio which includes all building materials and types of construction, and demonstrates exactly where and how savings can be made.

1, 2   Production and administration building „phg" in Augsburg, 2003

Do you prefer a particular building material for cost efficient building, or is every building material capable of providing the same cost savings when correctly utilised?
No, I do not prefer one particular building material. So many different factors have to be taken into account. What occurs to me on the spur of the moment, because it's very contemporary – is that three or four years ago it was very economical to span large spaces with steel; neither concrete nor timber structures could compete. Now, however, the tables have turned because all the steel is being bought up by the Chinese, and we are constructing in timber and concrete again. Market prices decide except when constructional requirements prevail.
Previously it was also more economical to span smaller spaces with concrete than steel. As the spans increase, the tendency to span with steel becomes greater. In our most recent large project, the Deka exhibition building, we used timber girders. Up until now, we have always implemented timber for large spans due to aesthetic rather than economic considerations.

Let us consider the showroom in Holland. Could you please explain briefly what you did in order to plan as economically as possible?
Gladly. The showroom is constructed in a standard way for Holland; that is that the building is mounted on piles obviating the need for other foundations.
Then we coordinated with the clients in order to develop a structural grid compatible with their exhibition system, thereby achieving an exceptionally economic product.
The resultant, clear built form is of a linear nature. The short ends are totally closed and the longitudinal sides are fully glazed (ill. 5). The glazing is subdivided into regular grid sizes, just large enough to appear generous, yet at the same time, the widths remain under the 2.3 metre limit beyond which the cost of such panels skyrockets.

Can you tell us something about the prefabrication process?
The entire building is completely prefabricated. The construction period comprised a total of three and a half weeks before the client could move in. Everything was wholly prefabricated in Germany; from the load-bearing panels, which were to be supported by steel girders, to complete room modules. The complete sanitary tract was delivered in a container and all services were installed; the only thing missing was the hand basin.

So you designed the sanitary tract and then it was only fabricated once?
We are more than happy to utilise systems but they often don't exist in the form required.
That is the problem with systems. There were probably sanitary cores in other dimensions available but not exactly the size we needed. Compacted layouts are unfortunately less flexible

Was it necessary to accept compromises in some areas of the showroom in order to realise cost efficient solutions? Or, conversely, did you deliberately select more expensive solutions when they were important for the architecture, and rather save in other areas?
The budget for the showroom was set at half a million Euros. In addition to the exhibition space, it was necessary to pro-

vide a training area, a customers' lounge and offices – a highly efficient prototype for numerous locations.

Of course, there was a certain amount of financial transfer back and forth. For example, we wanted to use a hardwood in the interiors. Now everything is in Larch, a very straightforward surface; robust rather than elegant.

As a result, however, we were able to realise our concept of the facade. We were quite sure that we wanted to do without cover strips and to apply a considerably more expensive version of glazing; it is after all an exhibition centre, a building that will be predominantly experienced from the outside. The priority was the appearance of an exceptionally smooth building without any form of profile and a minimum of construction. For all that is was necessary to accept certain compromises in the interior. The interesting thing about compromises is the possibility to experience another quality, one that is often well received by the users of the building. When flexibility is retained and the concept is allowed to develop - because it can develop in other ways – it is possible that totally new, unconsidered aspects can be created. New alternatives are always being found which, for minimal costs, can produce comparable quality. The skill lies in the ability to create an economical building which does not look cheap.

What was the reason for your involvement as investor with respect to cost efficiency? Was it possible to work more economically?

Our involvement as investor is a thing of the past. We are presently so well supplied with contracts that it is no longer necessary to work within that particular constellation.

Our experience as investor has definitely sharpened our sensibilities. One plans and builds more consciously; there is a huge difference between telling a client that the building will be 20% more expensive (which doesn't seem to be very much when one considers some colleagues) or if your office makes a profit of 15% but due to the 20% loss, you still have to find 5% of the building costs.

On the one side, we have developed an understanding for the investments of our clients. For the architect 10% of the construction cost is not a large sum, but for the person having to foot the bill it can be quite a lot. On the other side, the symbiosis of available funds and architectural quality was always present. This was always important to us, irrelevant of who was financing the construction. We always built for Architecture.

There was often the opportunity with turnkey constructions to guide money into the office account, by setting standards as low as they appeared in the brief. Many of our commercial competitors work on this basis – profit is priority. We, as architects, always wanted to create the best possible architecture. We were able to build well and economically, because I planned and operated these funds myself. When, for example, a contracted company constructed a building carcass 10% cheaper than quoted, that money was lost to architecture, and only the client profited. When, however, you retain control of the finances, it is then possible to purchase higher quality. Prerequisite are, of course, honourable intentions.

3, 4   Waldorf School in Augsburg, 2007
5   Showroom near Zevenbergen, 2005

Why did you give up building turnkey constructions?
This constellation can only function when costs and design are balanced. In addition to which, numerous administrative responsibilities come into play which were not conducive to creative thought. We certainly learnt a lot in the last ten years, but I must admit I sleep better and design better without that particular millstone around my neck.

Cost efficiency affects not only the building and its construction, but also the maintenance. What roles do future maintenance and operation costs play in the planning process?
Yes, construction costs are one thing, maintenance costs another and renovation costs yet another. That means, future costs resulting from cheap or shoddy work have no relation to the savings made at the time of construction. Unfortunately only few clients want to hear that. Over the last two years, however, a great amount of discussion has taken place, due to the exploding energy prices. Now there is hardly a building owner who is not interested in what his building will cost him when he has to manage it for 5, 10, 20 or 50 years. That can be measured primarily by energy consumption, and has a lot more to do with technical services. Of course, it is not only heating and running costs that have to be considered; but also what else can breakdown over a period of time. The discipline of Facility Management has become professionalized over the last ten years and everyone has slowly become aware of this.

Do architects also work as so-called advisors for energy consumption and operation of the building?
Absolutely. Consider this example. We built our first industrial hall with a long-term client who had built many large projects, at a time when no one was really interested in energy savings or heat retention, and we had a huge level of warm exhaust air under the level of the roof which was simply being blown out into the atmosphere. Now the entire company had to be converted to regenerative energy systems; a total of 35,000 m^2 that had to be converted in order not to import any more external energy. This idea has not developed slowly over the last ten years, rather it is the product of drastically inflating energy costs. The energy consumption in companies like these is exceptionally large and at some point in time thoughts turn to conservation. Now clients listen attentively where they would have closed the door in your face, only a year previously.

What new opportunities are there for architects now that consciousness is changing?
It is possible that we can communicate better, now that we represent the interface for all different topics related to construction. Previously it was simply accepted that the architect was responsible for all areas of expertise. Then, over decades, a change occurred based on the assumption that 20 specialists were necessary for a job previously carried out by 5 technical engineers, and the realm of the architect was limited to facade design. Now we are trying to return to that point where we are responsible for the coordination of all aspects of a project and are thereby capable of guaranteeing the overall result. This is considerably easier in relation to energy. Energy is not only the area of expertise of technical planners for heating, ventilation and sanitary installations, but also encompasses construction, thermal insulation technology, glazing, orientation and the spaces themselves.

You extended a school in Nepal. There you worked under totally different conditions, with the simplest of facilities; not only in the planning but probably in construction as well. What experiences did you have?
The Lophelling School was built in 1996 to house and educate 70 Tibetan refugee children. The most primitive materials were used. The school is located at the foot of the Annapurna Range at an altitude of 3,500 m and is dominated by extreme climatic conditions. We planned a new construction to be used as a prayer and dining hall, with kitchen and staff residences or relaxation rooms, as well as a wash house (ill. 8). It is even more apparent in Nepal than here,

6

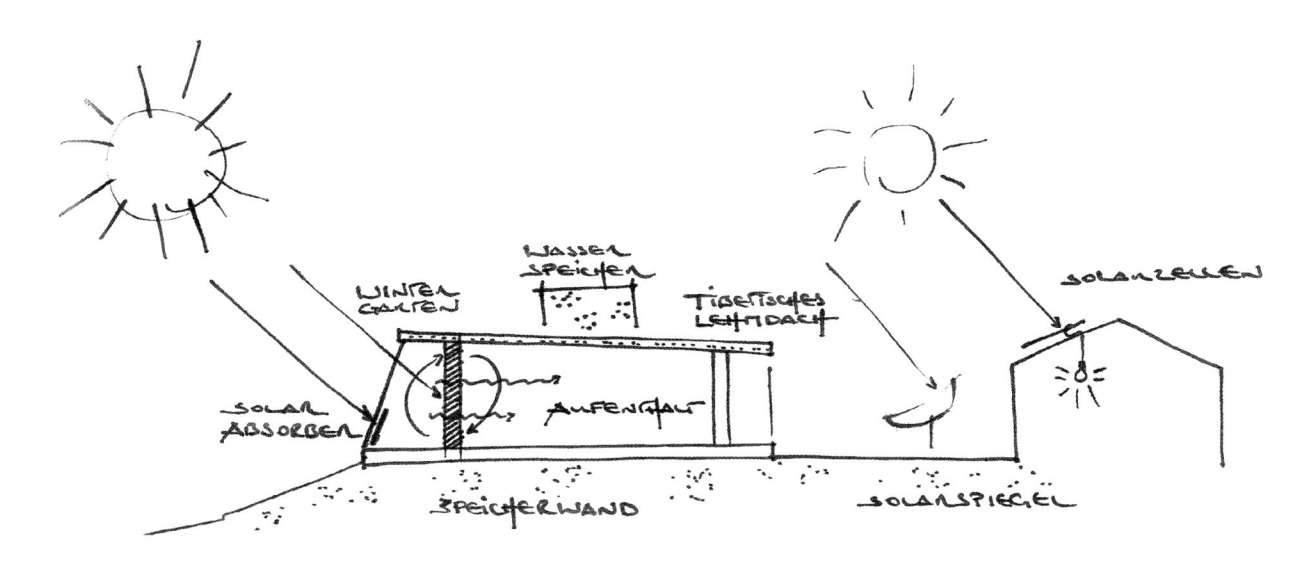

that many architects approach their projects without any concept for the real economy of the projects. Palaces are built there for the poorest. In many of the developmental projects I visited there, the necessary usable area measured a maximum of 50% of the total construction. The rest was given over to access and communication spaces, reception halls and spaces that were never even used. In our building there was not a single square metre designated as access space.

What can you tell us about the building process?
The production process in Nepal is the reverse of that here; material is very expensive, comparably speaking of course. The workforce, however, costs next to nothing.
That demands totally new thought processes and practical approaches. Here we purchase standardised windows in order to save money; there it is better to manufacture the same windows by hand on site. It can take three to four months before they are all completed.
Structural beams arrive as raw material – the tree trunks are delivered on the backs of the locals from the valley below, no one is allowed to fell trees up there anymore. Then they will be worked on site like here a hundred years ago. I found that very interesting as it offered new possibilities for creating something quite original, away from standardisation. The new glass house which we had built up there is 100% hand built, even the drip flashings were made on site. There are no half-finished building materials available there; they come from India via Kathmandu, require a 14 day foot march and are comparably expensive. We preferred supporting local industry and hand workers rather than some factory in India.
Handwork skills; working with metal, timber and stone, are gradually being lost due to industrialisation. This progress began a long time ago here.

How did the planning process develop with regard to construction?
We researched how the Tibetans built traditionally, before India started importing these stupid corrugated iron roofs which really do not belong in the landscape, but have unfortunately become an expression of affluence and a status symbol alike. Then we examined the construction of the Tibetan mud roofs; they are relatively primitive but also very cleverly constructed with layers of mud and waterproofing and built with a gentle fall. Eventually we decided to give the school building a mud roof. We earnt a great deal of ridicule over that! We were often asked why we wanted to save in that particular way. They were so impressed with the great glass house in front of the school, but couldn't understand our choice of construction for the roof. Finally it was not only a better solution financially but also climatically (ill. 7). Even at these altitudes it gets really hot in summer and cold in winter and, as you can imagine, there is an extreme difference between sleeping under a 40 to 50 cm thick mud roof and under one made of corrugated steel panels.
These were confrontations which reached back into the old traditions. One tries to discover how they think and where their values lie. We were fortunate to have two graduates on our staff that had each lived in Nepal for six months, and we therefore had good access to the people.
For example, I would gladly have done away with the ornamental carvings for financial reasons, but also for western, architectural reasons. But when you think about it, it is not kitsch here, it is hand work and is a part of their culture. It also encourages the hand work traditions and enhances the building in their eyes. We were prepared to compromise and were happy to accept this other aesthetic.

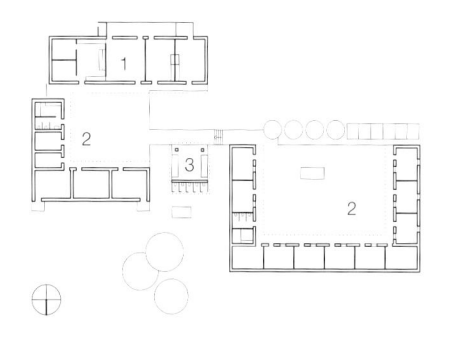

site plan
scale 1:1250
1 new building
2 existing building
3 wash house

6 Peutinger Secondary School in Augsburg, 2007
7 Lophelling School in Manang, Nepal, 2003;
  schematic sketches of energy concept
8 Lophelling School in Manang, Nepal, 2003

As far as material is concerned, we built almost everything in stone. These stones were carried out of the river bed and worked on site. That was about 80% of the construction. Timber is not really a local building material anymore because the government has forbidden anyone to fell trees in the region. That means that trees are no longer cut for building, only illegally for heating. Timber is brought from lower regions a couple of days walk away. The loam for the roofs was locally available and the timber windows were built on site. The glass panes, metal sheeting and cement were all brought up from the valley below. Of course we also had to lay a water pipe of a number of kilometres for the wash house.

And what about the foundations?
That's another story. We built a solid ring beam for the glass house to anchor it due to the wind suction. They have amazing storms in Annapurna in the afternoons when the wind comes up out of the Indian plains; the Kali Gandaki valley. The first sacks of cement which had been carried for 14 days up the mountain had all set when they arrived and the whole thing had to start again. Then, of course, we were continually being told by the locals "What do you want with cement up here, we've never done that!" Then we explained about the wind suction. The tribal elder just listened and then said "This does not happen in Nepal". He wasn't convinced, but we did it anyway.

Was the project in Nepal completed according to your wishes in the end?
Yes, we executed construction techniques which have found many followers – that was an ambition of ours. The school is located on the central tourism route; that means there are huge numbers of tourists passing on the Annapurna Trek. There are also a great number of guest lodges where visitors can spend the night. They consume enormous amounts of energy and contribute to the deforestation of the surrounding forests. My hope was that these hostel owners would take note of our construction techniques at some time and copy them – this has started to happen. The school has been finished for four years now and the feedback has been great; we are always being asked how we solved this or that problem. Particularly those elements that were most laughed at in the beginning have proven so popular that others are considering building in a similar manner in order to avoid the permanent race for firewood.

How were the finances organised for this project? Was it financed by donations or locally funded?
We financed this school ourselves. Our office was responsible for a large proportion of the funding and our business partners and clients assisted us with some very, very generous donations. It was possible to finance the project in a very short period of time.

Does anything occur to you that we could absorb from the planning and construction processes there?
It was confirmed there, that what we all search for, so-called simple construction, is actually a very, very complex form of construction – in detail and in the production. Strangely enough it seems to always turn around and become complicated. That is a cycle that I hope to understand one day. Why is it that this simplicity, that we all strive for is, at best, realised in the form. When we examine the constructions closely,

when we look at the details, they are usually a long way away from being simple. Zumthor builds highly aesthetic buildings which, when reduced to the last detail, are amazingly complex, even complicated.

What I believe, or what I hope, we brought home from Nepal is a particular simplicity of building, or even more casual buildings, which allow more scope and flexibility. Not every corner must be finished with an 8 mm shadow joint, one should allow the surfaces more freedom, and perhaps the individual elements could then reveal more of their inner selves.

I took these thoughts with me from Behnisch. He practised this philosophy in another way; with an individuality that we in our society attempt to produce in a way that has nothing to do with freedom. There is usually an enormous financial or planning effort concealed.

Where I think all or most of us want to end up, is building like the Mexican architect Barragán, who built extremely strong and extremely simple structures which were also rather imprecise in detail.

There are, of course, many contributing factors. The clients in Middle and South America are totally different from here. They don't even need building permission, they just start building. That is a totally different framework which demonstrates why it is possible to build things that are much, much simpler, and in the end appear simpler than here.

9 Lophelling School in Manang, 2003; complete construction
10 Bearers transporting building material for the Lophelling School
11 Lophelling School in Manang, 2003; pupils cleaning the windows of the new glass house

Studio in Madrid

Architects: Abalos & Herreros, Madrid

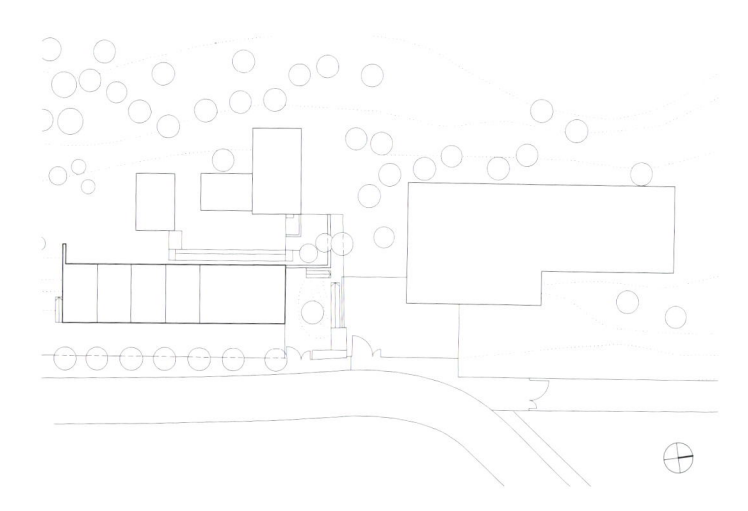

Using the natural features

This studio, located on a narrow strip of building land on the outskirts of Madrid, has the rear elevation entirely buried into the site. The load-bearing structure, which consists of windowless reinforced concrete walls, is spanned by a series of steel beams, with trussed girders in the areas of the skylights.

Homogenous appearance

The building envelope is wholly clad with polycarbonate sheeting. In order to maintain the constancy of the facade, the doors are also executed in plastic. This enables the internal life of the building to be visible from the outside, albeit blurred and indistinct, while the windows and skylights glow in the surrounding darkness. Where no insulating concrete wall is located behind the facade envelope, the architects have employed double layers of corrugated panels; the intermediate layer of air acts as insulation.
The translucent polycarbonate cladding ensures even, indirect lighting internally, a decisive factor for the client, who is a painter. Plasterboard walls, white painted steel elements and terrazzo flooring create a neutral, restrained, though pleasant atmosphere, despite the lack of visual interaction with the outside world.

Natural sun shading with vegetation

The roof of the studio is partially covered by those plants which are capable of flourishing in this harsh, dry climate. They also spread across the half-buried rear elevation.

site plan
scale 1:750
section
floor plan
scale 1:250

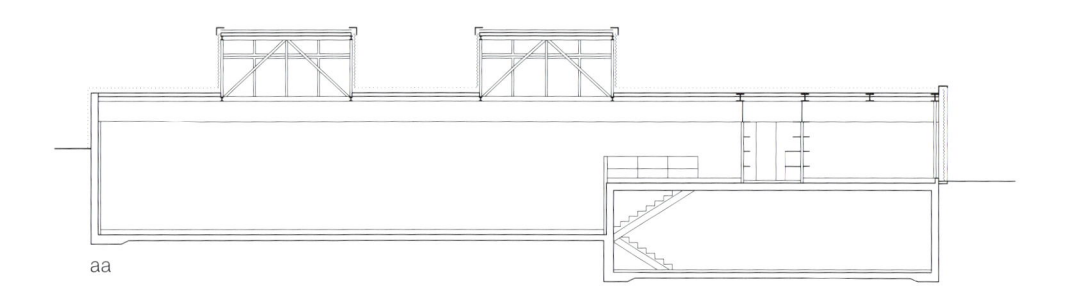

aa

Project details:
Usage: Studio
Construction: reinforced concrete
Total construction cost: 500,000 Euros
Cost per m² residential area: 1,524 Euros
Total storey area: 302 m²
Total usable area: 328 m²
Built site area: 404 m²
Total site area: 2,058 m²
Date of construction: 2002
Period of construction: 1999–2002

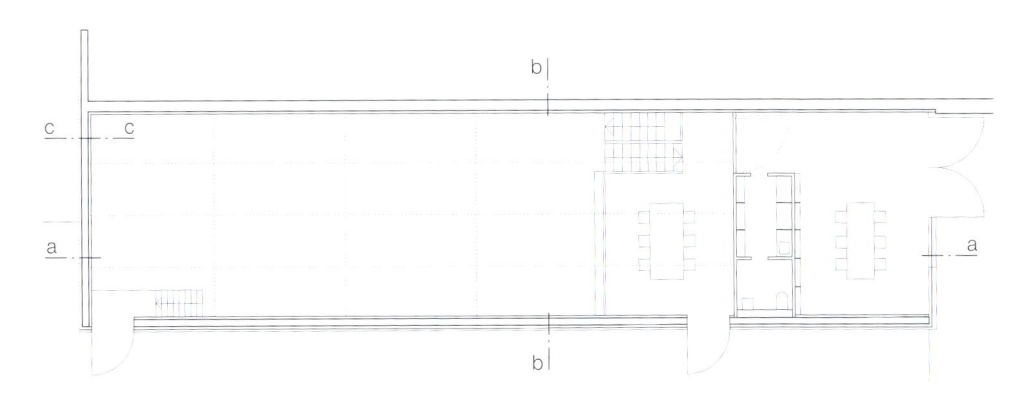

vertical section
horizontal section through door
scale 1:20

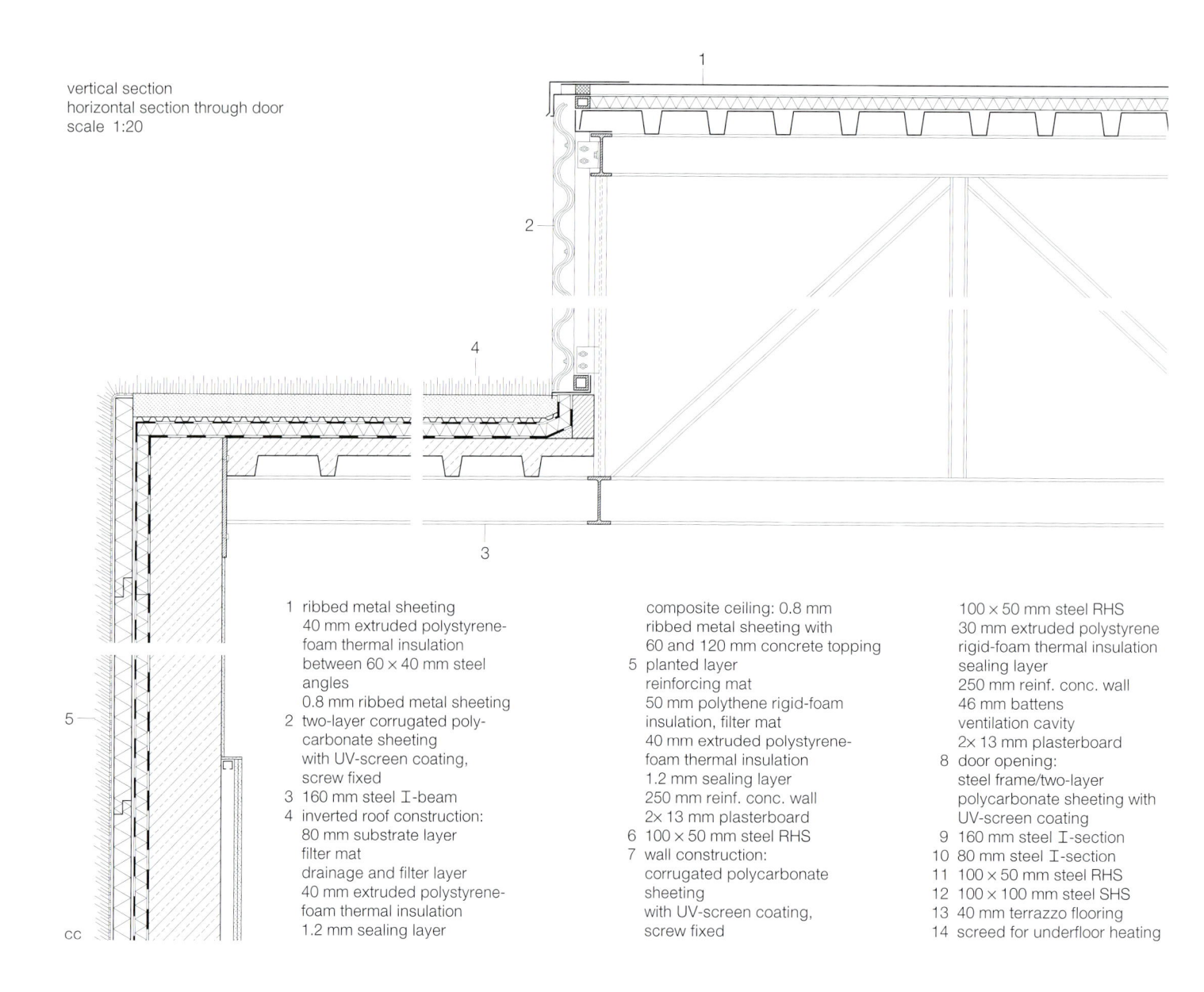

1 ribbed metal sheeting
  40 mm extruded polystyrene-foam thermal insulation
  between 60 × 40 mm steel angles
  0.8 mm ribbed metal sheeting
2 two-layer corrugated polycarbonate sheeting
  with UV-screen coating, screw fixed
3 160 mm steel I-beam
4 inverted roof construction:
  80 mm substrate layer
  filter mat
  drainage and filter layer
  40 mm extruded polystyrene-foam thermal insulation
  1.2 mm sealing layer
  composite ceiling: 0.8 mm ribbed metal sheeting with
  60 and 120 mm concrete topping
5 planted layer
  reinforcing mat
  50 mm polythene rigid-foam insulation, filter mat
  40 mm extruded polystyrene-foam thermal insulation
  1.2 mm sealing layer
  250 mm reinf. conc. wall
  2× 13 mm plasterboard
6 100 × 50 mm steel RHS
7 wall construction:
  corrugated polycarbonate sheeting
  with UV-screen coating, screw fixed
  100 × 50 mm steel RHS
  30 mm extruded polystyrene rigid-foam thermal insulation
  sealing layer
  250 mm reinf. conc. wall
  46 mm battens
  ventilation cavity
  2× 13 mm plasterboard
8 door opening:
  steel frame/two-layer polycarbonate sheeting with UV-screen coating
9 160 mm steel I-section
10 80 mm steel I-section
11 100 × 50 mm steel RHS
12 100 × 100 mm steel SHS
13 40 mm terrazzo flooring
14 screed for underfloor heating

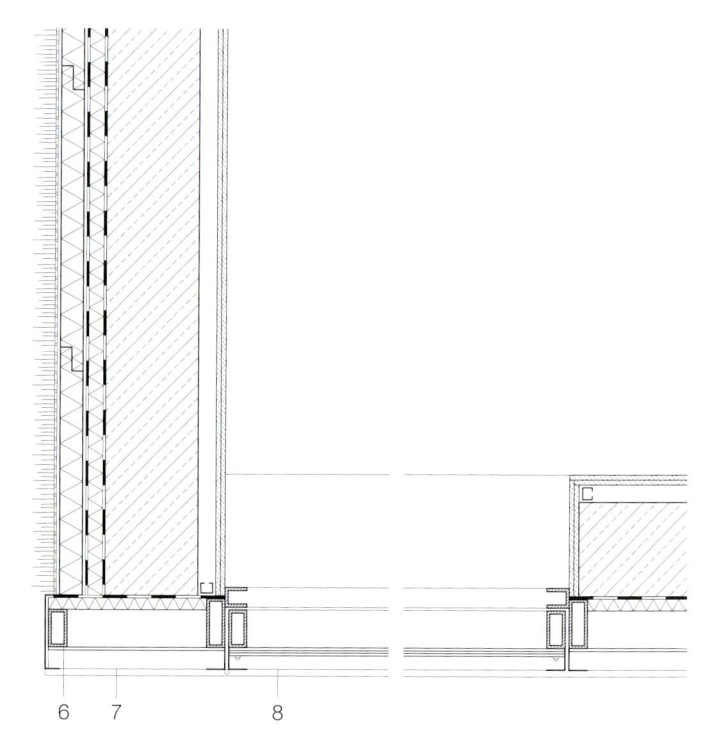

bb

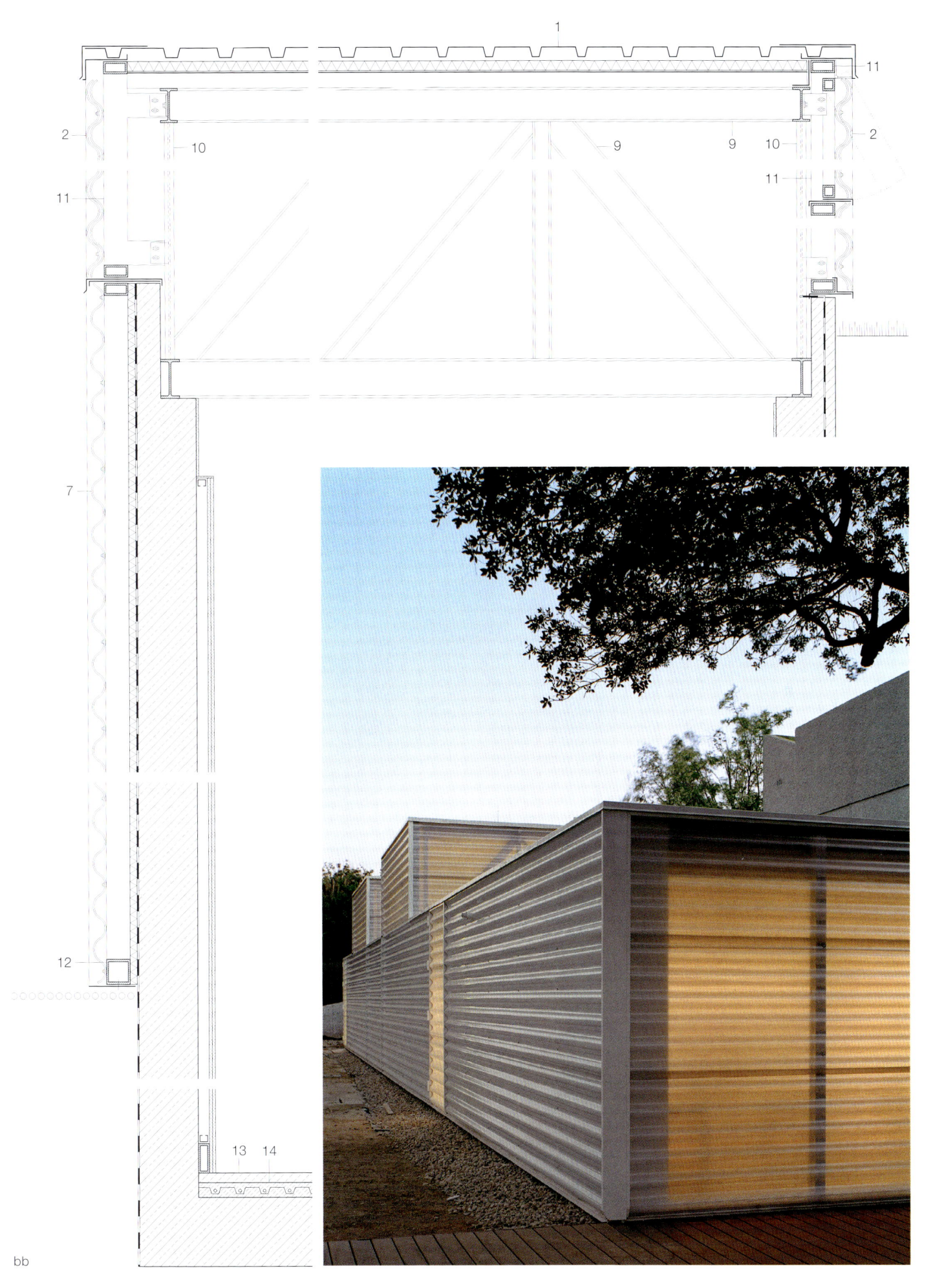

Pharmacy and Medical Practice in Plancher-Bas

Architects: Rachel Amiot and Vincent Lombard, Besançon

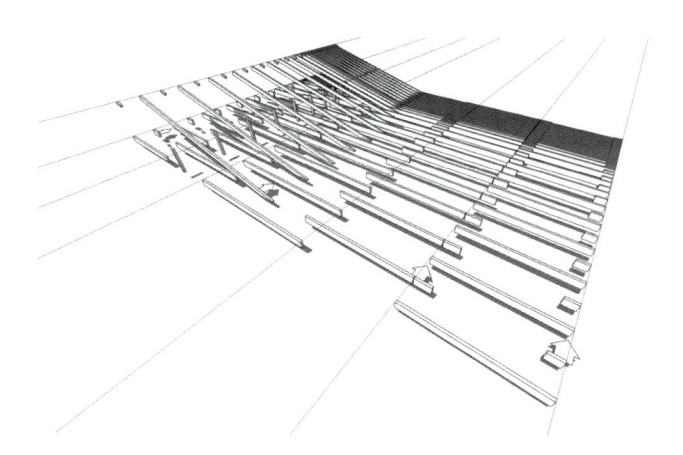

**Short construction time due to parallel site works
Light-weight prefabricated timber construction
Filigree roof constructed of standard sections**

The soaring roofs of this modest health centre in Plancher-Bas, in north-eastern France, only allow brief glimpses of the delicate roof structure concealed beneath. These deconstructed timber plates, developed by the structural engineer Jean Luc Sandoz, contribute immensely to the economy of the project as a whole (p. 10–11).

Clear structure and refined design

A pharmacy is accommodated in the lower storey of this clear-cut two storey cube, while the upper level is given over to a medical practice. Due to the sloping nature of the corner site, both facilities benefit from direct access to the ground level. The interactive character of the building is enhanced by the generously proportioned communication zones incorporated into the design.

Differentiation of the facades and horizontal articulation of the building is achieved by the use of a restricted number of different materials. Punctuating the otherwise closed facade of the concrete plinth level, large format shop windows offer views into the pharmacy, while the upper level is enveloped in a layer of louvers. These vertical larch timber elements protect the privacy of the horizontal windows to the medical examination rooms. Individual louvers can be swung up in the vicinity of the opening window elements in order to provide uninterrupted views into the surrounding countryside.

Efficient timber construction techniques

The separation of trades, legible in both material selection and storey differentiation, made it possible for the building to be completely constructed within six months. Even as the plinth storey was being concreted, the carpenters were prefabricating the entire timber construction for the upper storey. After completion of the structural built carcass, the finished timber-framed walls were assembled. The walls carry the loads from the roof structure, which spans up to seven metres in some areas and is also of timber construction. Elements of standard cross-sectional dimensions were offset butted and screw fixed to form large constructional slabs which required only 0.13 m^3 of timber per square metre of slab. Apart from the direct material and cost savings, this also reduced structural loads and thereby the dimensions of other load-bearing elements. The upwardly thrusting roof forms were based on the same construction principle and are braced in the horizontal planes. These roofs satisfy the pitch requirements of the municipal building authorities, provide additional illumination – in conjunction with the loggias – into the internal spaces and characterise the timber ceilings.

Project details:
Usage: retail (lower floor), medical practice (upper floor)
Construction cost: 670,000 Euros
Cost per m^2 usable area: 1,196 Euros
Cost per m^2 roof area: 115 Euros
Usable area: 560 m^2
Total built area: 615 m^2
Total internal volume: 1,940 m^3
Site area: 970 m^2
Construction period: six months
Date of construction: 2004

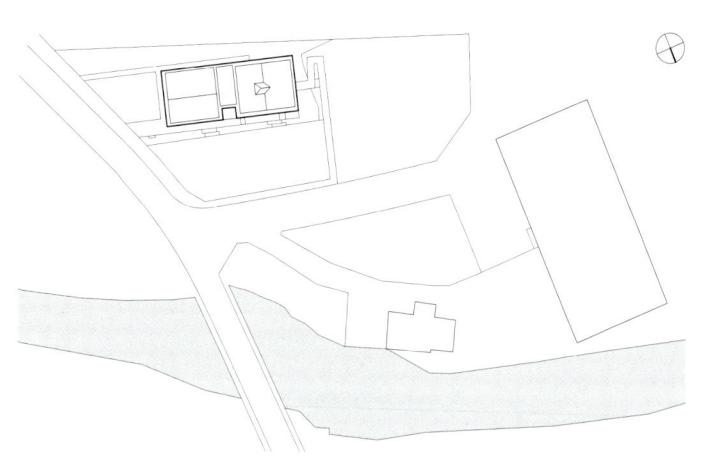

aa bb

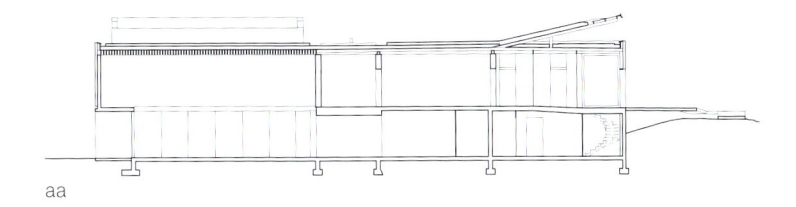

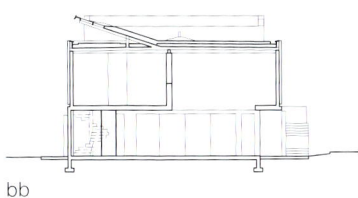

site plan
scale 1:2000
roof construction isometric

floor plans · sections
scale 1:400

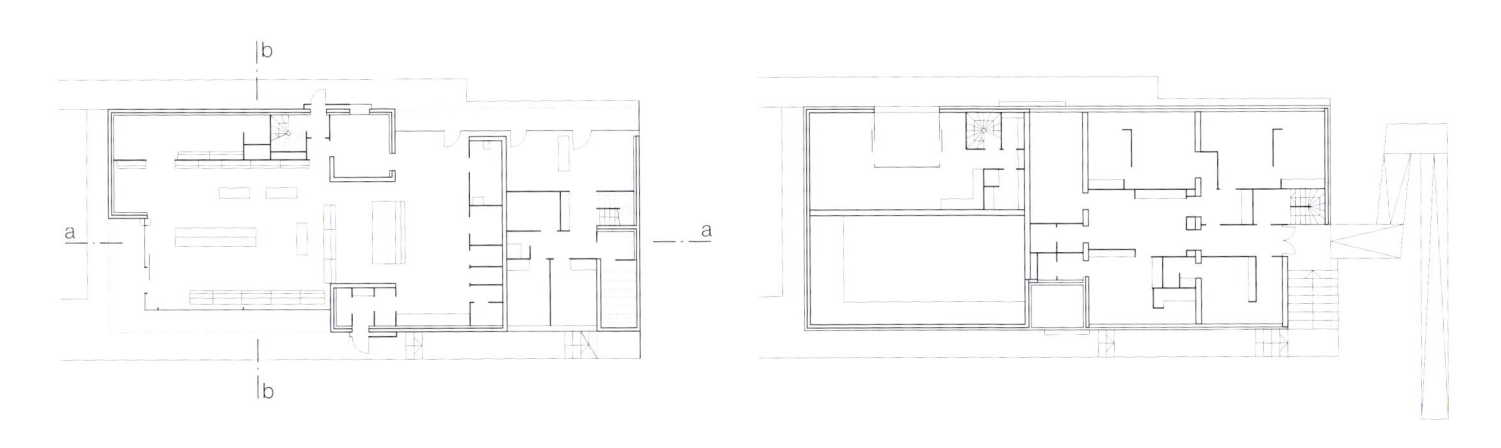

PHARMACIE

30

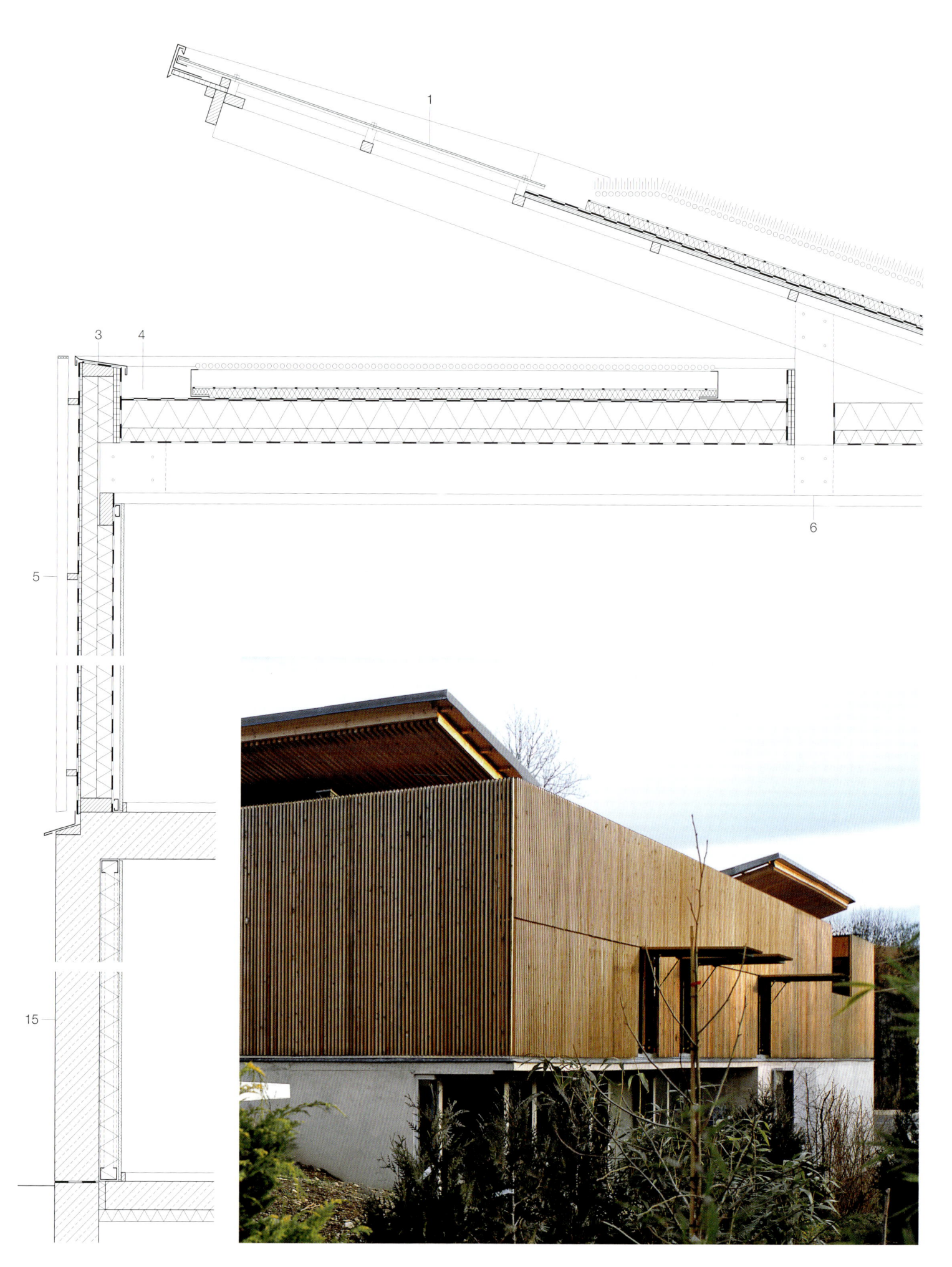

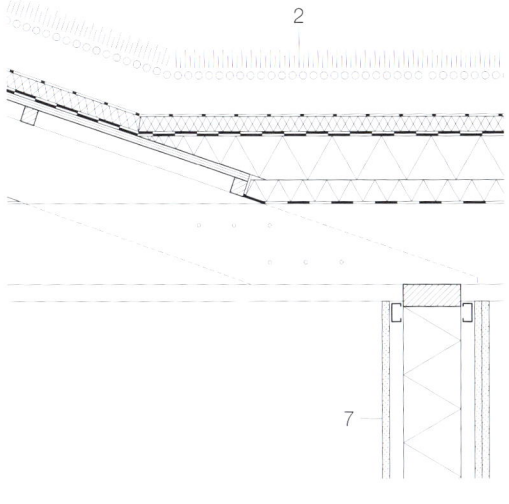

vertical sections • horizontal section
scale 1:20

1 corrugated transparent polycarbonate panels
2 50 mm vegetation, 100 mm substrate
  40 mm filter mat, drainage layer
  20 mm impermeable layer
  120 mm thermal insulation
  60 mm thermal and acoustic insulation
  vapour barrier, 50 × 220 mm fir roof system,
  offset butted and screw fixed
3 1 mm zinc parapet flashing
4 gutter
5 wall construction: 25 × 45 mm larch battens
  27 × 45 mm fir sub-construction
  black polyester matting
  12 mm oriented-strand-board
  140 mm mineral wool thermal insulation
  between 140 × 60 mm fir studs
  vapour barrier, sub-construction
  12.5 mm plasterboard
6 50 × 180 mm fir column
7 wall construction:
  60 × 160 mm fir studs clad both sides with
  12,5 mm plasterboard, 160 mm insulation
8 window construction:
  double-glazing in timber frame
9 shutters, manually operated
  25 × 45 mm larch battens
10 Ø 12 mm steel safety barrier
11 floor construction:
  200 mm reinforced concrete, sanded
12 24 × 40 mm fir skirting
13 zinc window sill
14 double glazing in steel frame
15 wall construction:
  200 mm reinforced concrete
  100 mm thermal insulation
  12,5 mm plasterboard
16 floor construction:
  120 mm reinforced concrete, sanded
  50 mm thermal insulation

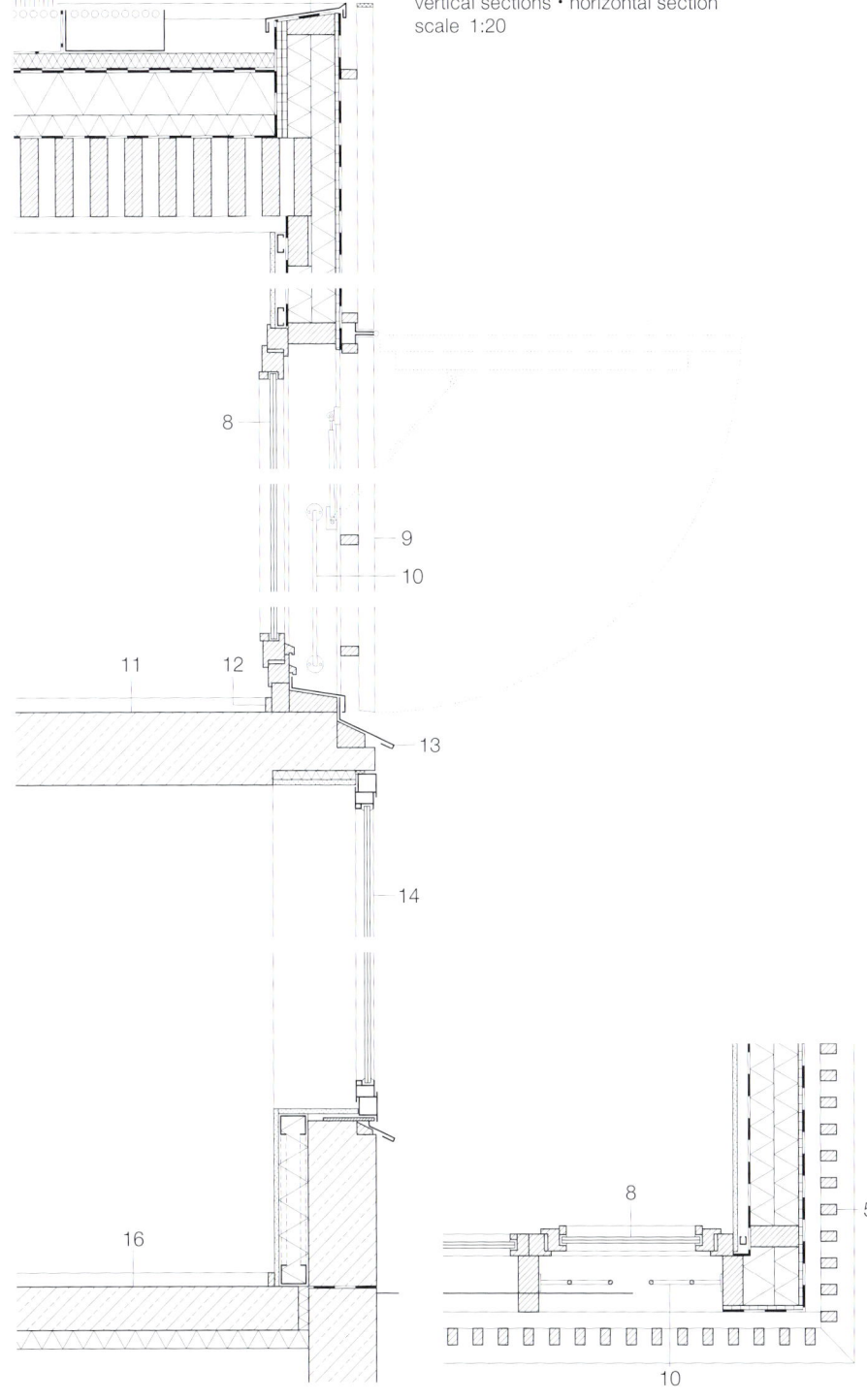

Static Lightness in Timber

Jean-Luc Sandoz, CBS-CBT, Lausanne

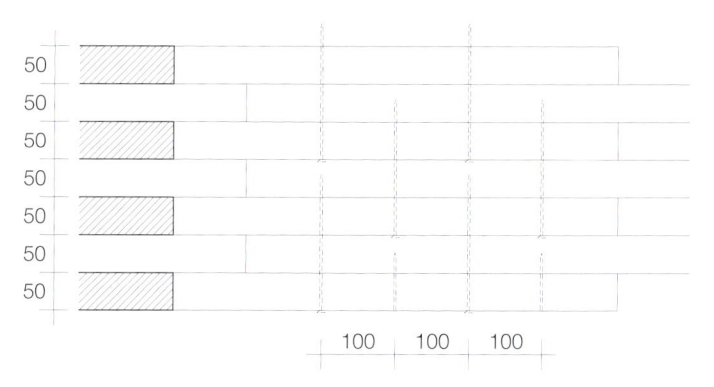

2

The O'Portune slab system

The O'Portune slab system, developed by Mr. Sandoz, was selected for the pharmacy and medical practice in Plancher-Bas. The O'Portune slab represents an optimisation of the stacked plank system. The planks are again fixed together in a vertical position, in an offset manner in order to achieve a greater structural depth and rigidity with the same quantity of timber. The individual planks are screwed rather than nailed together, whereby each screw passes through up to three shear joints. Due to the dimensions of these slabs, effective sound insulation and a fire resistance of 30 minutes (even up to 60 minutes with increased cross-section of elements) are attainable without the need for additional measures.

Constructional variations according to span

For medium spans of between 5 and 8 m, O'Portune beams are used instead of continuous slabs. The beams occupy a half or a third of the overall floor area. This is referred to as a "Solivium" slab system and functions like a conventional beam structure with integrated planar elements; oriented-strand board, plywood, etc. The visual contrast between the structural elements and the in-fill panels can also offer aesthetically pleasing design alternatives.

For spans of between 9 and 11 m, the O'Portune slab system would cover the entire floor area and be capable of bearing loads of up to 300 kg/m^2. The individual planks, measuring from 40 × 80 mm up to 80 × 220 mm, are of solid timber with dovetail joints. Panels of oriented-strand board or plywood provide high performance floor slabs capable of transferring horizontal loads. The open channels on the underside of the slab considerably improve acoustic performance.

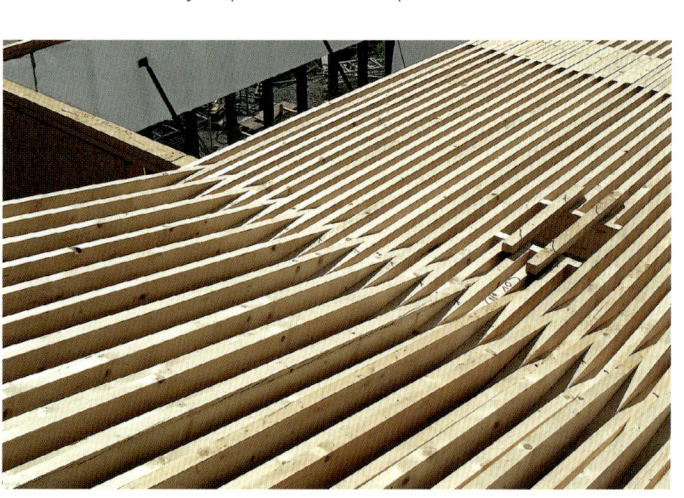

3

To bridge even greater spans, of up to 18 m, or to bear larger loads, the timber floor can be combined with concrete to create a form of composite construction. The system is based upon the use of timber as reinforcement in the tension zone, with a concrete filling designed to bear compression loads.

Prefabricated in panel widths of 1.2 metres, the slab systems can be treated with either clear varnish or coloured paint, as desired. This exceptionally high level of prefabrication enables areas of up to 500 m^2 to be laid in one day, while the relatively small weight, of 100 kg/m^2 for a span of up to 10 metres, allows considerable reductions in the load-bearing structure. The aesthetic treatment of the slabs can be controlled by the use of various plank dimensions and profiles, timber selection and coating alternatives. The inherently posi-

1

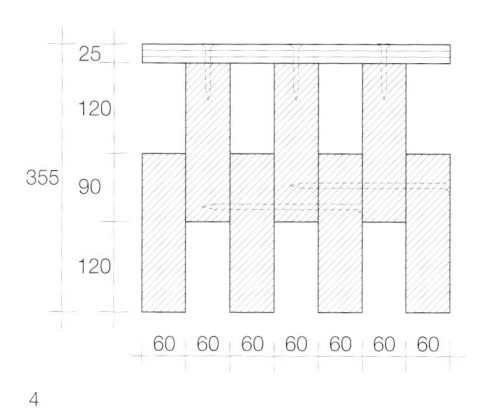

tive acoustics of such a system can be further enhanced through the application of additional sound insulating materials. An additional advantage is the simple integration of technical installations; electrical, air-conditioning, or data-transfer, into the slab system.
For the same price as a traditional beam system, this timber slab system offers many advantages:
- high level of prefabrication
- swift assembly
- static lightness
- simple integration of technical services
- effective acoustics
- flexible, contemporary architectural alternatives

The roof slab system of the pharmacy and medical practice

For this project in Vosges it was possible to do away with the upper layer of off-set planks, due to the relatively minimal span of only seven metres, at the specific request of the architect. This allowed the planks to be off-set butted and screw-fixed. The final weight of the slab was kept down to 50 kg/m² compared with that of a standard O'Portune slab of 100 kg/m². Only 1/3–1/5 of the weight loading of a comparable concrete slab was applied to the load-bearing system, thus enabling the dimensions of the structural members to be dramatically reduced.
A dynamic soffit was achieved by off-setting the planks, in addition to the effective acoustic qualities which were further enhanced by applying sound insulation panels (rockwool with selected coloured matting) to the planks.
It was a simple matter to realise the soaring roof forms which give the project its own particular character while allowing light to penetrate into the structure, and additionally demonstrate this particular roof system to both external and internal observers.
This character of openness and light is continued into the facade treatment; where both the privacy and intimacy of the internal spaces are protected while simultaneously providing light and views. The vertical panelling, of 25 × 45 mm larch louvers, is set in front of the actual building envelope, animating and enlivening the facade, and relieving the solidity of the cubic construction. Additionally, the windows are more easily cleaned – a positive effect when considering maintenance costs. This building "skin" opens like a shutter and allows views into the surrounding countryside. Although larch timber is often used in external situations, the cut edges were additionally capped with PVC covers (see detail section pp. 30–31).
Due to the high level of prefabrication; timber framed walling, slabs and facade, it was possible to complete the construction within six months and to achieve significant cost-savings of over 30%.

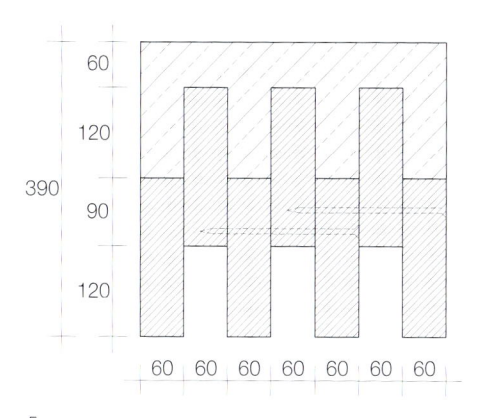

1 Roof construction pharmacy
2 Pharmacy structural system
3 Window shutter pharmacy
4 O'Portune system section
5 D-Dalle system section
6 O'Portune slab soffit
7 D-Dalle experimental sample

Mobile House in England

Architects: Mae Architects, London

Project details:
Usage: mobile dwelling
Construction: timber framework (prototype)
steel framework
Internal ceiling height: 2.5 m
Total construction cost: 195,000 Euros
Cost per m² usable area: 1,950 Euros
Total internal volume: 215 m³
Total built area: 86 m²
Usable area: 100 m²
Period of construction: 12 weeks
Date of construction: 2002

**Mobile dwelling with high standard of living
Two prefabricated building segments
Timber frame construction
Individuality**

The "m-house" combines the advantages of a mobile, temporary dwelling with the standard of a high quality loft. The location is flexible – whether as holiday home in the wilderness, roomy dwelling on a roof top or building extension in a rear courtyard.

The two prefabricated segments, each measuring 17 × 3 m, are delivered to the desired location 12 weeks after the order is placed, and assembled along the longitudinal axis in a single day. Following connection to the local infrastructure, the house is ready for occupation. In addition to a large, open living and dining area with views through ceiling-high timber sliding windows, the dwelling offers two bedrooms with integrated bed and robe systems, a bathroom, WC, and additional utility room. The interior walls are continuously clad with pale birch plywood, which creates a contrast with the black linoleum floor in the open living space and the dark brick surround behind the wood-burning oven.

Energy and acoustic requirements have been met, and a highly comfortable internal atmosphere achieved, with the incorporation of insulation in the prefabricated system and the provision of underfloor heating in the living area. The client can choose from a variety of options for the facade treatment, ranging from corrugated aluminium cladding to more sedate cedar boarding, according to personal taste.

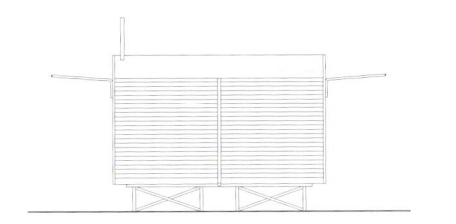

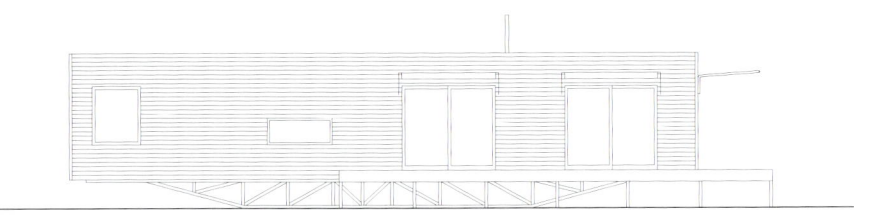

floor plan • sections • elevations
scale 1:200

A module 1
B module 2
1 entrance
2 living room
3 veranda
4 services
5 bathroom/WC
6 bedroom

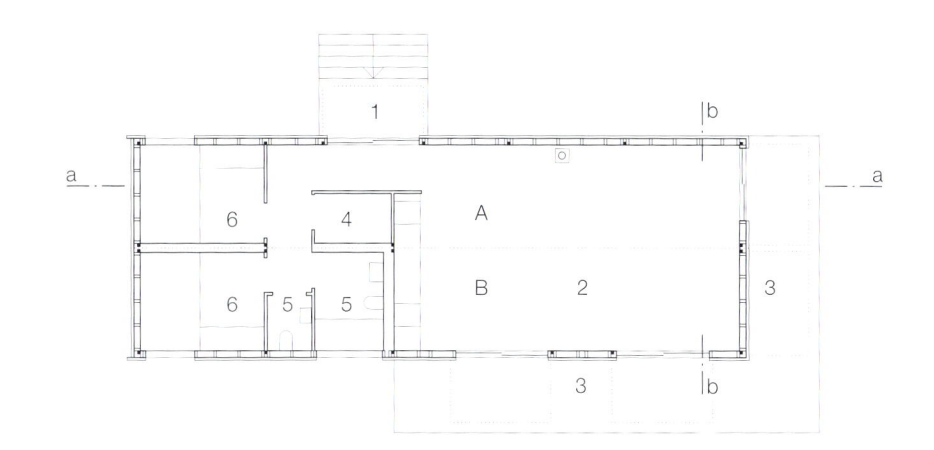

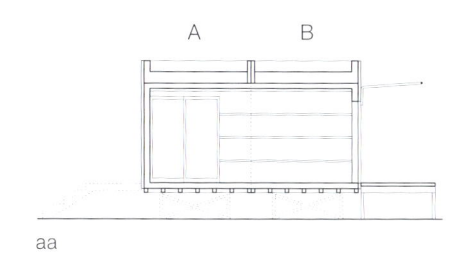

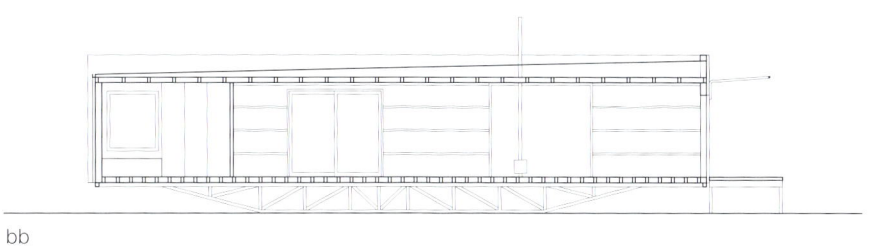

aa

bb

sections scale 1:10

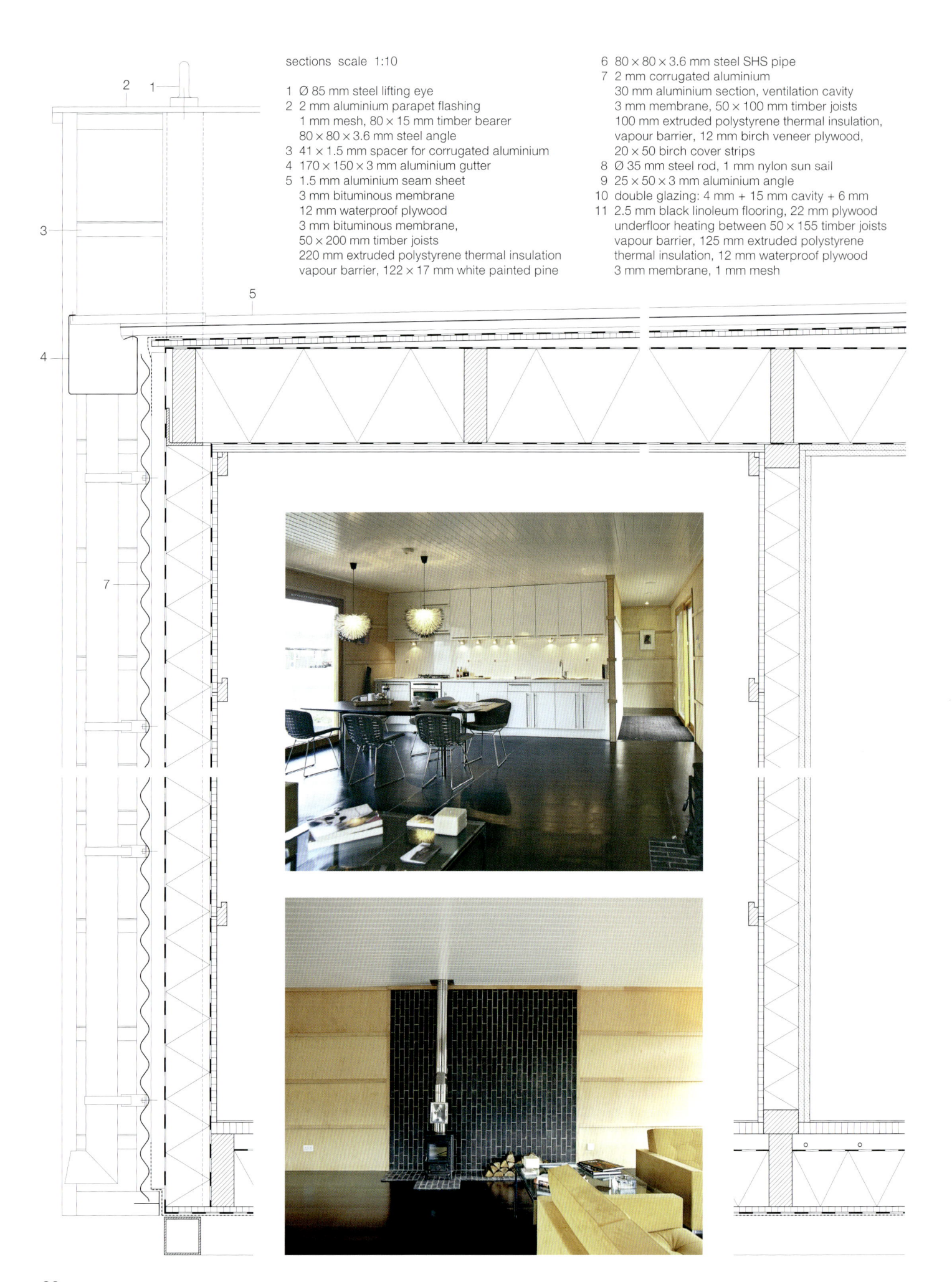

1. Ø 85 mm steel lifting eye
2. 2 mm aluminium parapet flashing
   1 mm mesh, 80 × 15 mm timber bearer
   80 × 80 × 3.6 mm steel angle
3. 41 × 1.5 mm spacer for corrugated aluminium
4. 170 × 150 × 3 mm aluminium gutter
5. 1.5 mm aluminium seam sheet
   3 mm bituminous membrane
   12 mm waterproof plywood
   3 mm bituminous membrane,
   50 × 200 mm timber joists
   220 mm extruded polystyrene thermal insulation
   vapour barrier, 122 × 17 mm white painted pine
6. 80 × 80 × 3.6 mm steel SHS pipe
7. 2 mm corrugated aluminium
   30 mm aluminium section, ventilation cavity
   3 mm membrane, 50 × 100 mm timber joists
   100 mm extruded polystyrene thermal insulation,
   vapour barrier, 12 mm birch veneer plywood,
   20 × 50 birch cover strips
8. Ø 35 mm steel rod, 1 mm nylon sun sail
9. 25 × 50 × 3 mm aluminium angle
10. double glazing: 4 mm + 15 mm cavity + 6 mm
11. 2.5 mm black linoleum flooring, 22 mm plywood
    underfloor heating between 50 × 155 timber joists
    vapour barrier, 125 mm extruded polystyrene
    thermal insulation, 12 mm waterproof plywood
    3 mm membrane, 1 mm mesh

37

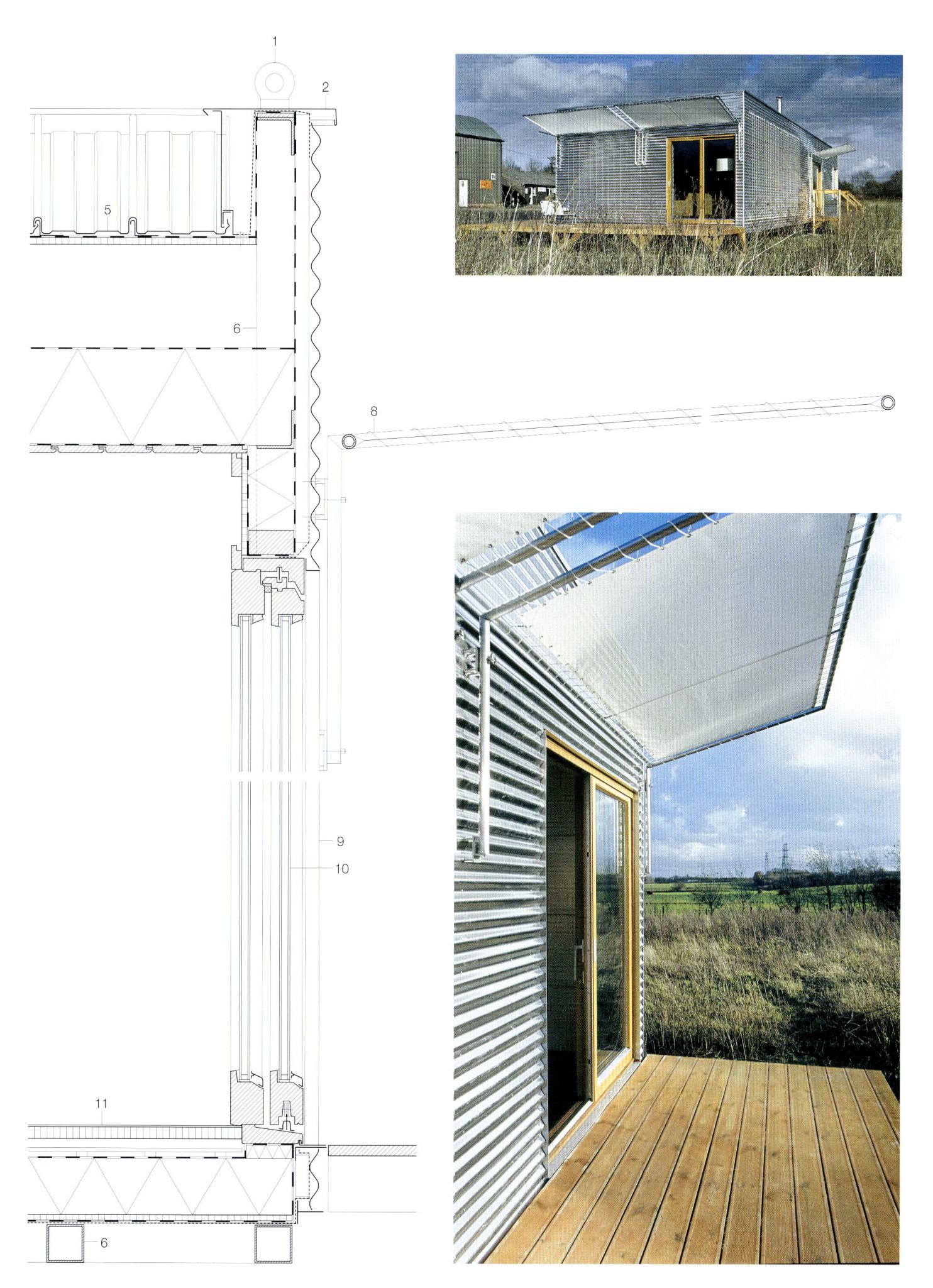

Prefabricated House from Denmark

Architects: ONV arkitekter, Vanløse

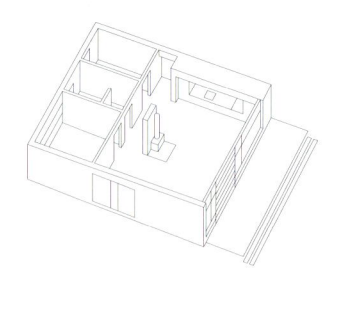

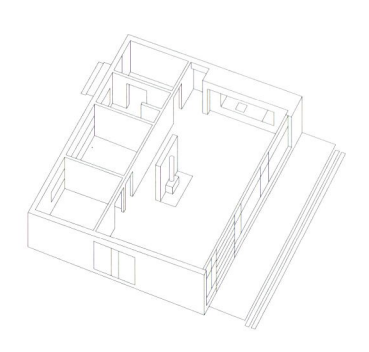

A B

Cost savings through prefabrication
Modules of various sizes
Smallest module completed prior to delivery

While selecting furniture from a catalogue is not such an unusual thing to do, it is now also possible to order the corresponding house. In contrast to comparable concepts this house, developed by the Danish architectural firm ONV, has attained high levels of design quality and flexibility. The minimalist residence is available in six basic variations.
Due to the high degree of prefabrication, the structure can be manufactured a relatively low cost. The smallest version is produced entirely in the factory, and can be delivered to site by road, while the larger versions are made up of two to four modules. On site these prefabricated modules are set on strip footings, then the roof sheeting is sealed, the skylights installed, and finally the connections cleaned. The timber framed external walls are clad in Siberian larch and the interior walls with gypsum fibreboard.
In order to keep circulation space to a minimum, all versions are planned around a large communal space which accommodates living and dining areas, and an open kitchen which can be connected to a covered terrace as desired. The internal and external spaces are linked to each other via extensive glazing and the number of additional rooms remains flexible. Depending upon the selected module combination, the final living area can vary from 60 to 170 square metres. These modular construction techniques also allow simple, economical extensions to be viable at any point in the future.
Intelligent detailing enables the use of high quality ash parquet and natural stone flooring within a modest budget.

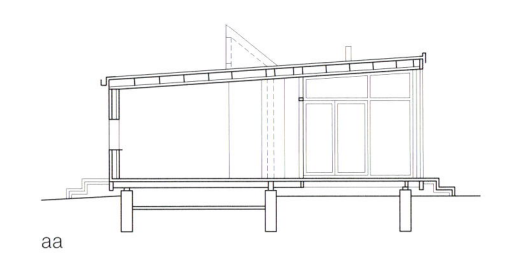

aa

C   E   F

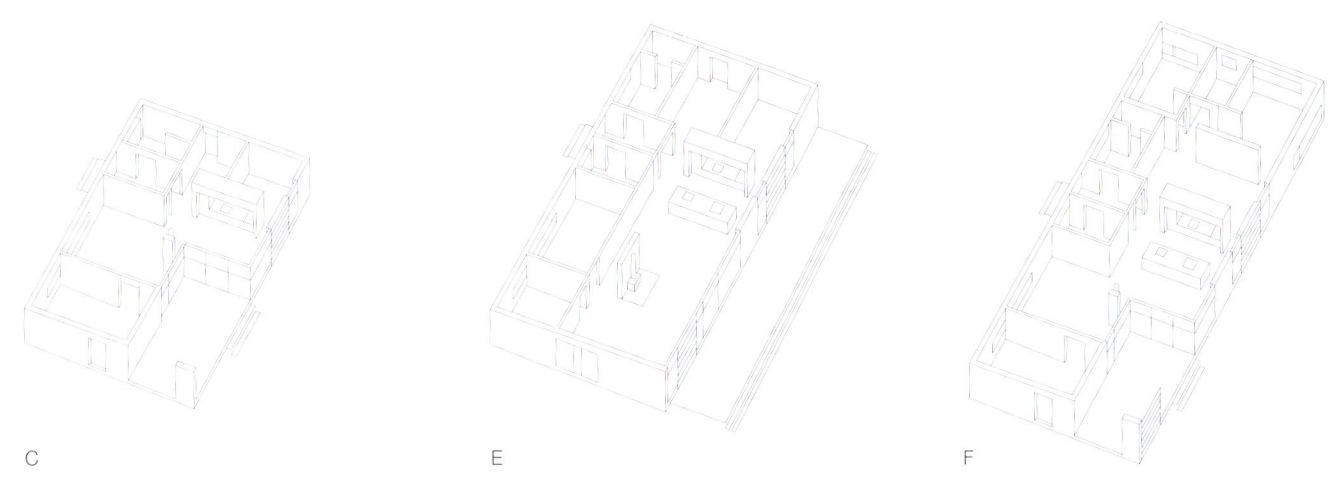

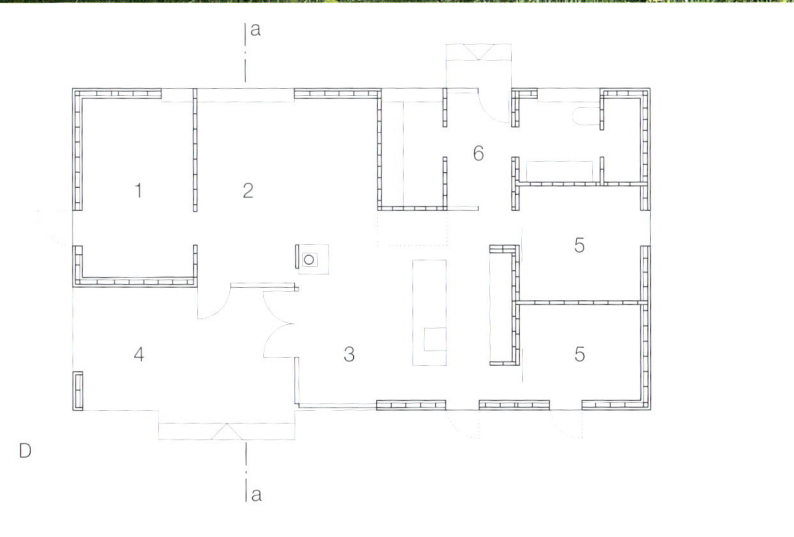

Project details:
Usage: Prefabricated single
                          family house
Construction: timber framework
Cost per m² usable area: 1,400 Euros
Total floor area: 60 m² – 169 m²
Construction time: 8 – 12 weeks

floor plan · section   Axonometric drawings
scale 1:200   of the available versions

1 bedroom   version A   60 m²
2 living   version B   86 m²
3 dining   version C   103 m²
4 covered terrace   version D   134 m²
5 room   version E   138 m²
6 entrance   version F   169 m²

39

horizontal section • vertical section
scale 1:20

1 tongue-and-groove boarding
  22 mm larch, transparent coated
  22 mm battens with ventilated cavity
  12 mm plywood
  145 × 45 mm post-and-rail facade
  145 mm mineral-wool insulation
  vapour barrier, 45 mm insulation
  between 45 × 45 battens,
  15 mm gypsum fibreboard
2 120 mm larch frame double glazing:
  12 mm + 6 mm cavity + 12 mm
3 15 mm gypsum fibreboard
  70 mm thermal insulation between
  70 × 45 mm timber studs
  15 mm gypsum fibreboard
4 120 × 120 mm steel angle column
5 65 × 233 mm laminated timber beam
6 2-ply bituminous sheeting, welded
  on site, 15 mm plywood
  195 mm mineral-wool insulation
  between 195 × 45 mm rafters,
  vapour barrier 45 mm insulation
  between 45 × 45 mm battens
  15 mm gypsum fibreboard
7 14 mm parquet
  vapour barrier, 15 mm plywood
  195 mm mineral-wool insulation
  3 mm synthetic panel
8 timber bearing member with
  ventilation outlet
9 100 mm reinforced concrete slab,
  with drainage opening
10 segment connection:
  90 × 233 mm laminated timber
  header, with 50 mm joints
11 sliding element terrace:
  30 × 30 mm galvanized steel angle
  45 × 25 mm larch louvers,
  transparent coated
12 300 mm reinforced concrete strip
  foundation, filled on site

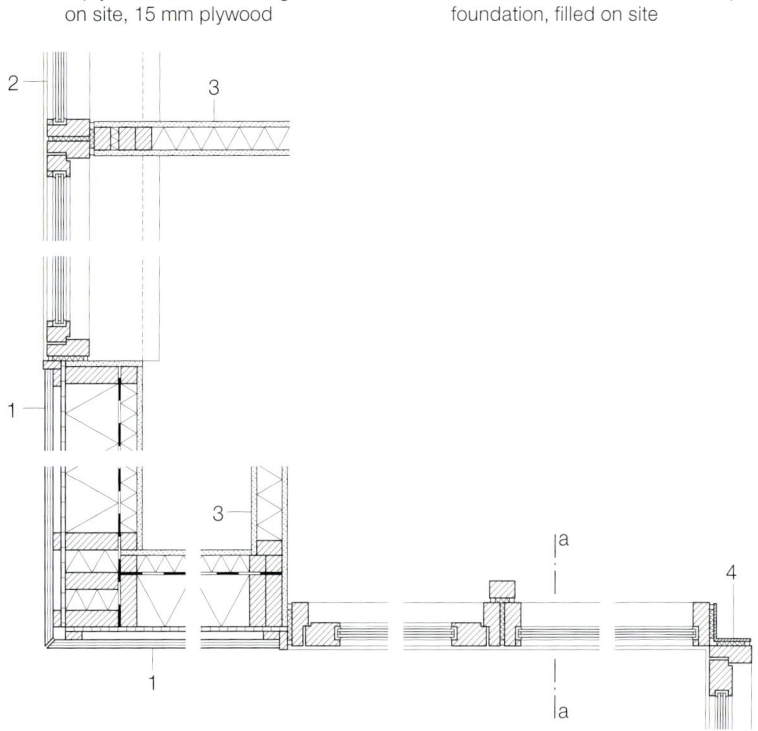

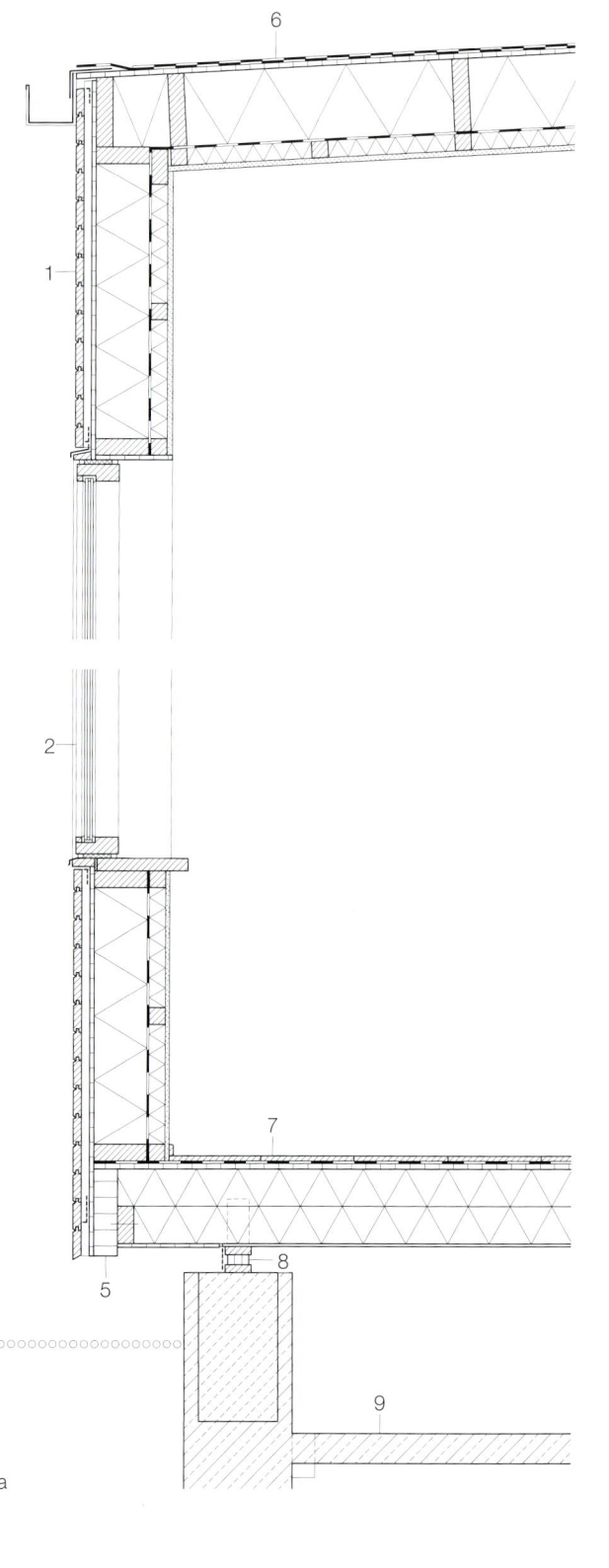

41

Straw House in Eschenz

Architect: Felix Jerusalem, Zurich

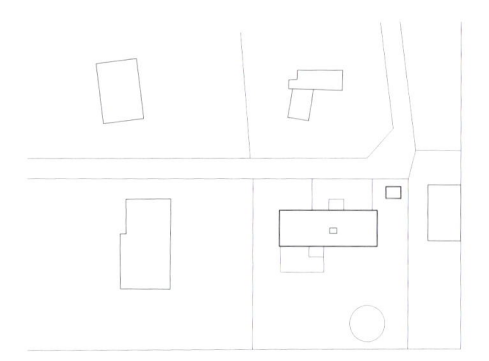

Pile construction with concrete core

On the southern bank of Lake Constance in a small Swiss village, this simple, but well-conceived house, which was built on a tight budget, accommodates a family of four. In view of the wet subsoil conditions, the structure was raised above the ground on piles. The only section that extends down into the earth is the solid concrete core, which accommodates sanitary facilities, a galley kitchen, a cloakroom and a cellar. The cellar space is accessible via a trapdoor in the floor. The core element also articulates the elongated house into two sections. These comprise the two children's rooms at the southern end; and to the north, the living area and parent's bedroom, with a study in the gallery above which takes advantage of the additional height afforded by the roof slope.

Compressed straw-fibre panel construction

The internal spaces have the charm of a refined carcass structure. The flooring consists of sealed screeds, service runs are visibly mounted, and the walls have been left partially untreated. Of special interest, however, is the form of construction. Except for the core, it was built of compressed straw-fibre elements, a material that is both pollution-free and recyclable. The floors, ceilings and walls consist of sandwich elements all based on the same principle: the outer layer of highly compressed straw-fibre panels performs the load-bearing function, while the lightweight straw-fibre filling acts as thermal insulation. All elements were prefabricated and assembled on site. The entire structure was completed in only four months.

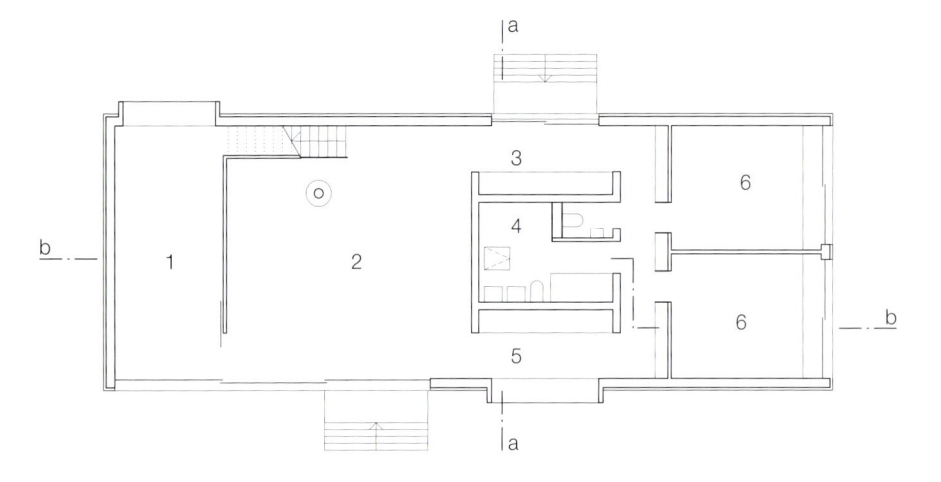

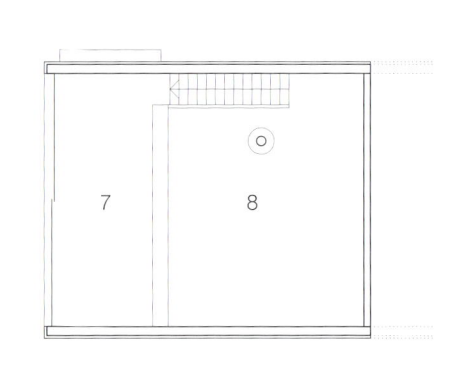

site plan
scale 1:1500

floor plans · sections
scale 1:200

1 bedroom
2 living
3 entrance
4 bathroom
5 kitchen
6 children's room
7 gallery
8 void

Project details:
Usage:	Private residence
Construction:	Timber/straw
Internal ceiling height:	2.3–4.9 m
Total cost of construction:	426,738 Euros
Cost per m² residential area:	2,463 Euros
Total internal volume:	844 m³
Total built area:	173 m²
Usable area:	140 m²
Total site area:	1,077 m²
Date of construction:	2005
Period of construction:	3 months

aa

bb

43

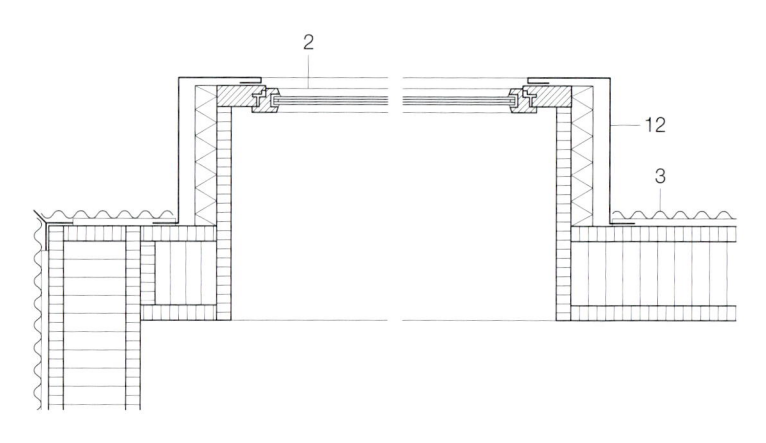

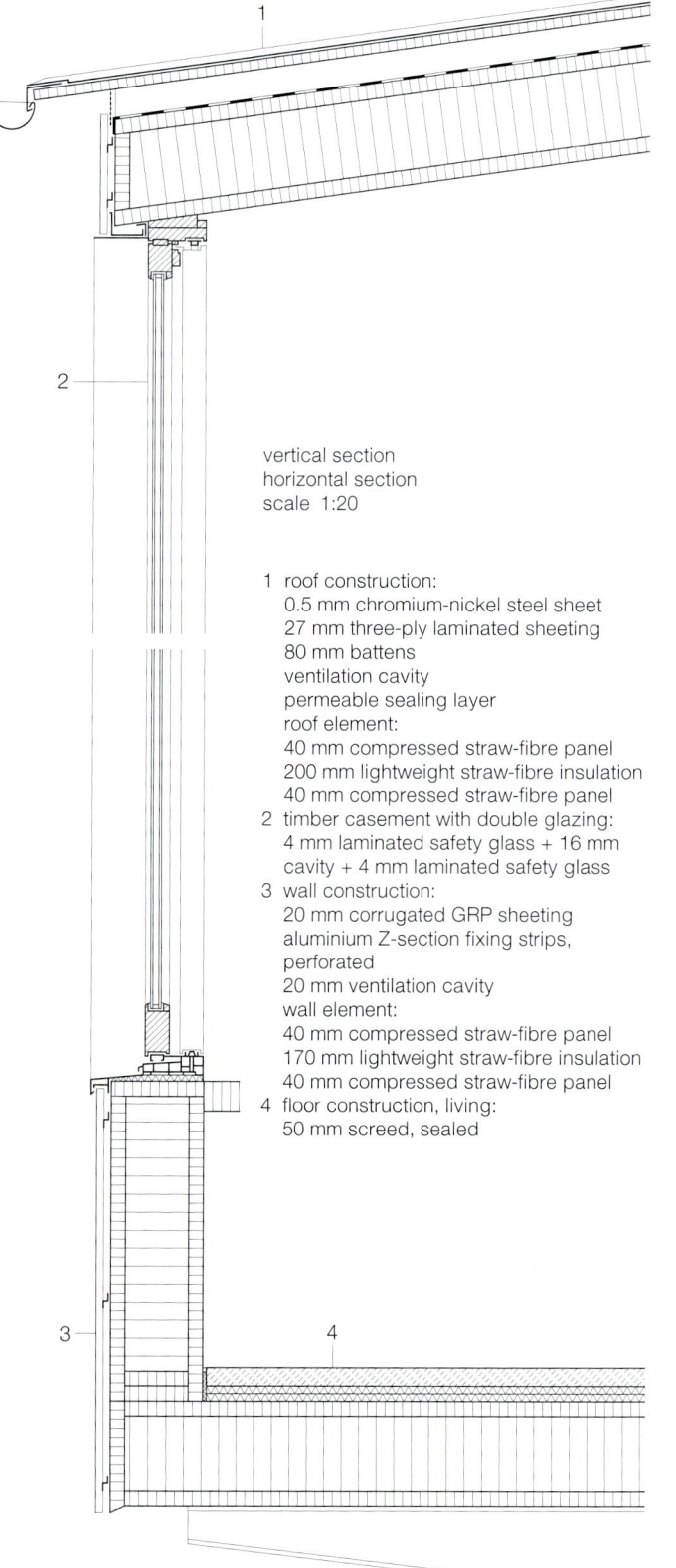

vertical section
horizontal section
scale 1:20

1 roof construction:
   0.5 mm chromium-nickel steel sheet
   27 mm three-ply laminated sheeting
   80 mm battens
   ventilation cavity
   permeable sealing layer
   roof element:
   40 mm compressed straw-fibre panel
   200 mm lightweight straw-fibre insulation
   40 mm compressed straw-fibre panel
2 timber casement with double glazing:
   4 mm laminated safety glass + 16 mm
   cavity + 4 mm laminated safety glass
3 wall construction:
   20 mm corrugated GRP sheeting
   aluminium Z-section fixing strips,
   perforated
   20 mm ventilation cavity
   wall element:
   40 mm compressed straw-fibre panel
   170 mm lightweight straw-fibre insulation
   40 mm compressed straw-fibre panel
4 floor construction, living:
   50 mm screed, sealed

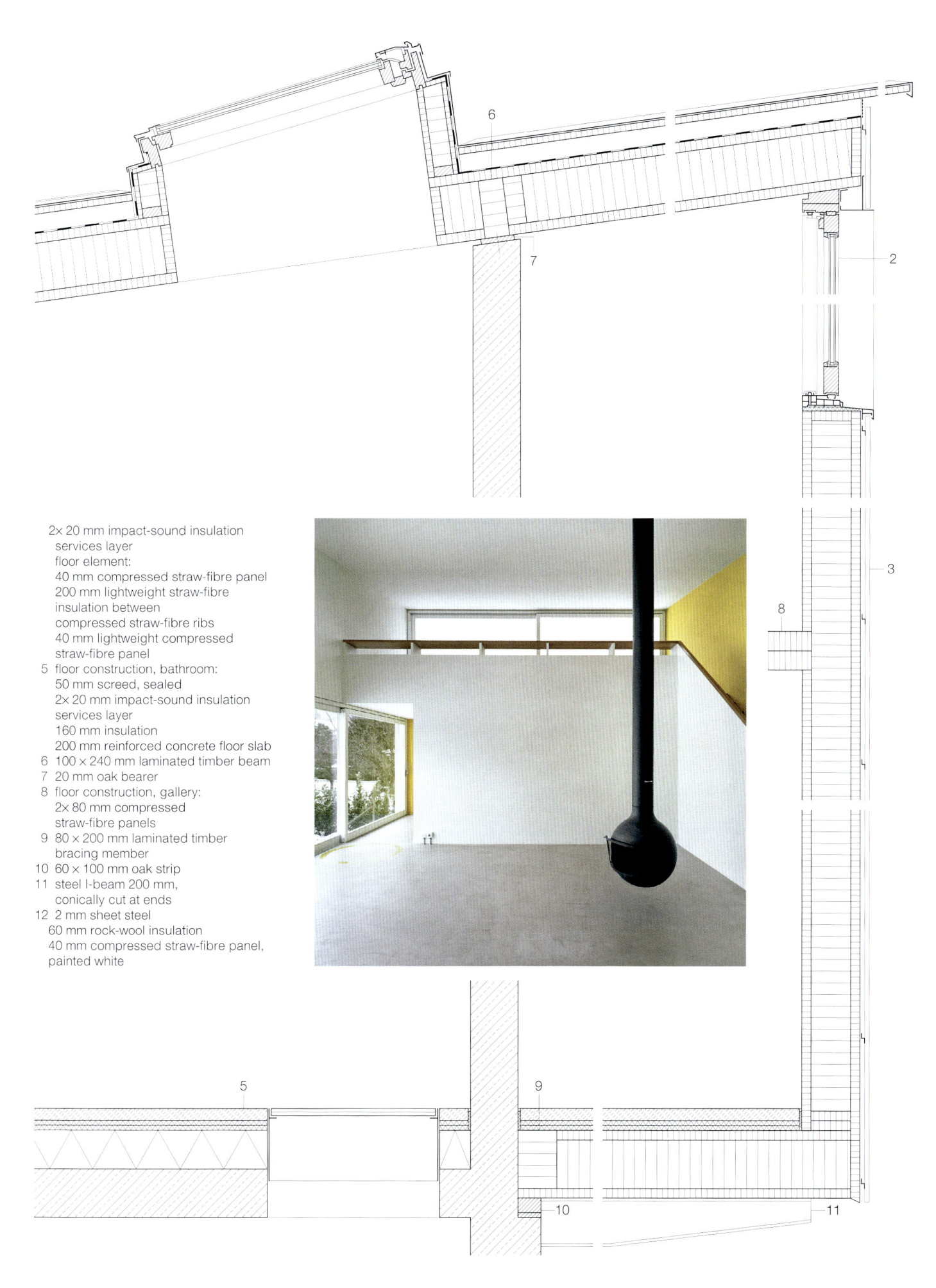

2× 20 mm impact-sound insulation
services layer
floor element:
40 mm compressed straw-fibre panel
200 mm lightweight straw-fibre
insulation between
compressed straw-fibre ribs
40 mm lightweight compressed
straw-fibre panel
5 floor construction, bathroom:
50 mm screed, sealed
2× 20 mm impact-sound insulation
services layer
160 mm insulation
200 mm reinforced concrete floor slab
6 100 × 240 mm laminated timber beam
7 20 mm oak bearer
8 floor construction, gallery:
2× 80 mm compressed
straw-fibre panels
9 80 × 200 mm laminated timber
bracing member
10 60 × 100 mm oak strip
11 steel I-beam 200 mm,
conically cut at ends
12 2 mm sheet steel
60 mm rock-wool insulation
40 mm compressed straw-fibre panel,
painted white

House in Shimane

Architects: Sambuichi Architects, Hiroshima

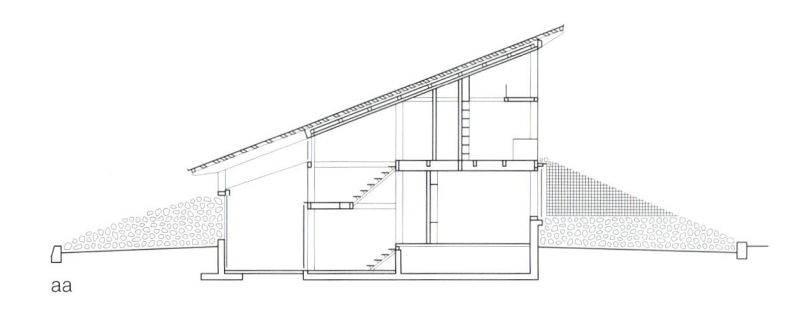

Adaptation of location, form and internal organisation to the extreme climatic conditions

The house is situated in west Japan, in a region that experiences marked climatic contrasts between seasons. In summer, there is subtropical heat, while winter is characterized by icy temperatures and deep snow. The architect has attempted to create a concept which compensates for these climatic extremes. Although surrounded by rice paddies, the house is buried into a mound of stone rubble that does not absorb moisture. Only a comparatively small area of the house is exposed to the environment; the glazed roof is oriented to the south to maximize solar gains in winter, while air cavities form a thermal buffer between internal and external temperatures.

Exploiting natural resources

When buried in snow, the stones shield the ground floor against wind and cold, while in summer they maintain a temperature balance. It is then that the timber structure comes into its own. By opening the entrance gates and the glazed sliding elements, the entire house is cross-ventilated from the patio. The closed and transparent roof areas which allow convective climate control also enrich the interior spaces with an interesting play of light and shadow.

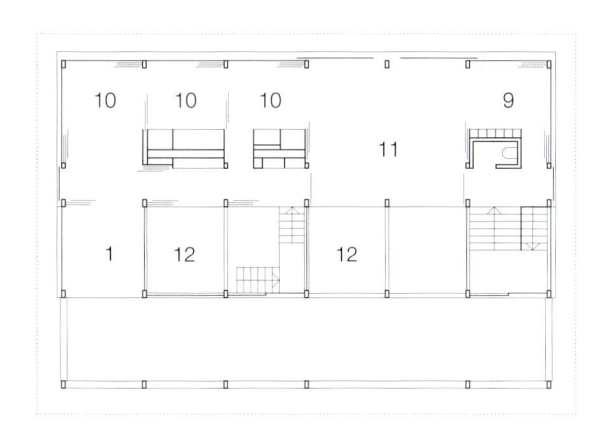

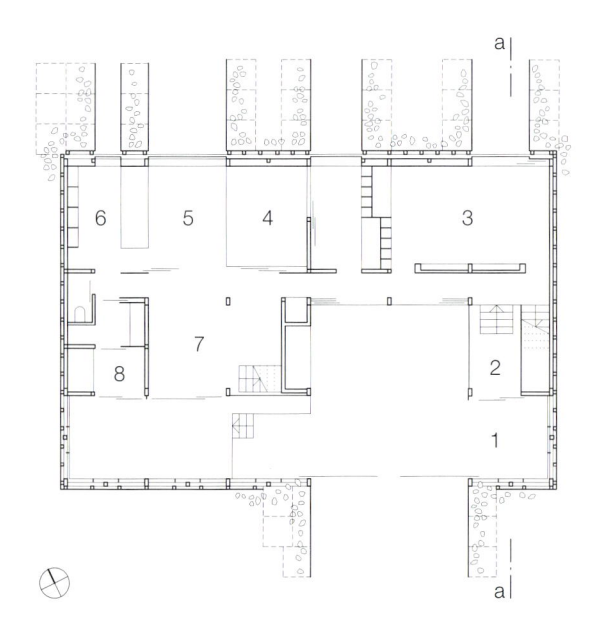

section · floor plans
scale 1:250

1 patio
2 entrance
3 bedroom
4 living
5 dining
6 kitchen
7 hall
8 bathroom
9 study
10 guestroom
11 terrace
12 void

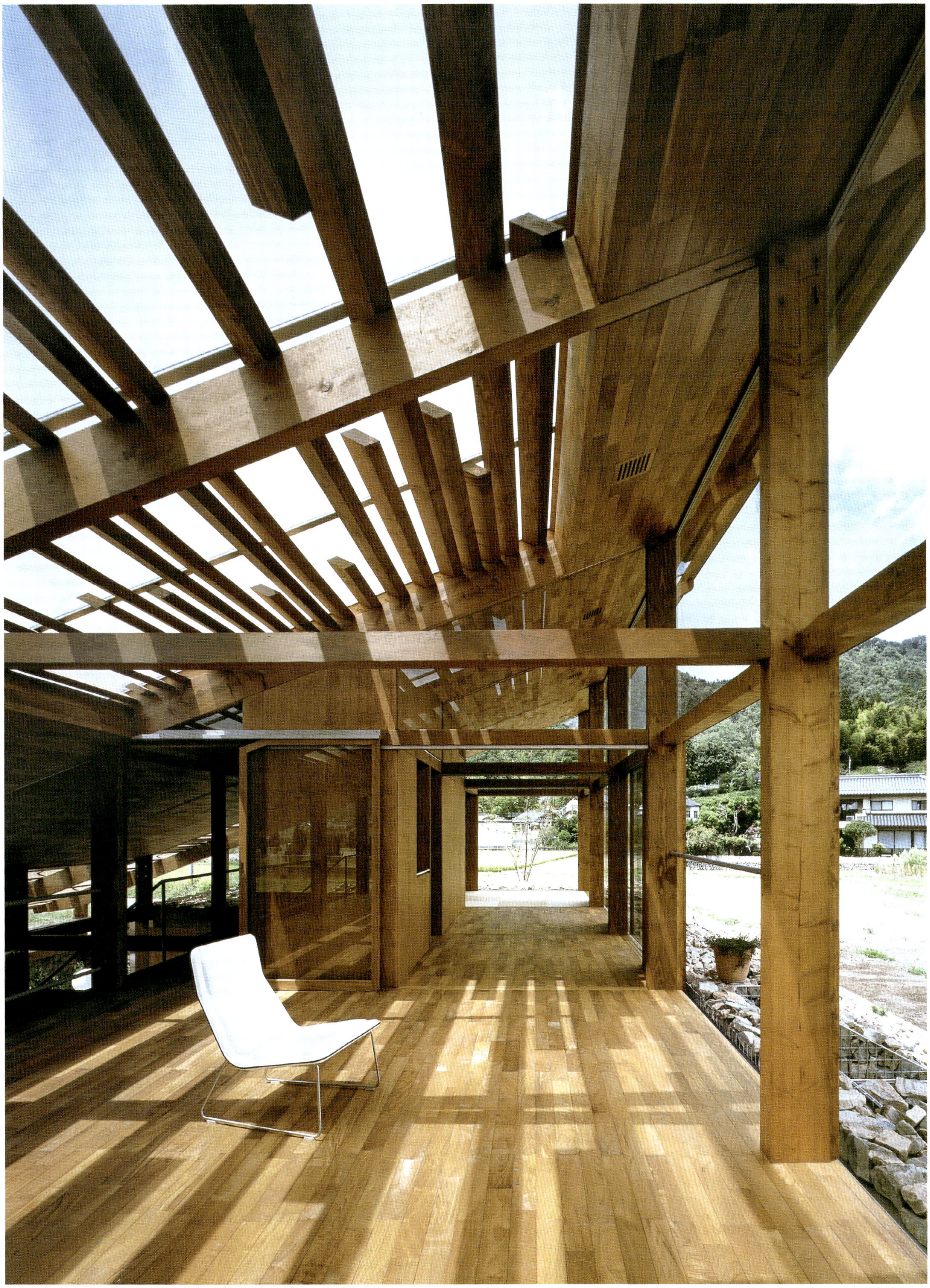

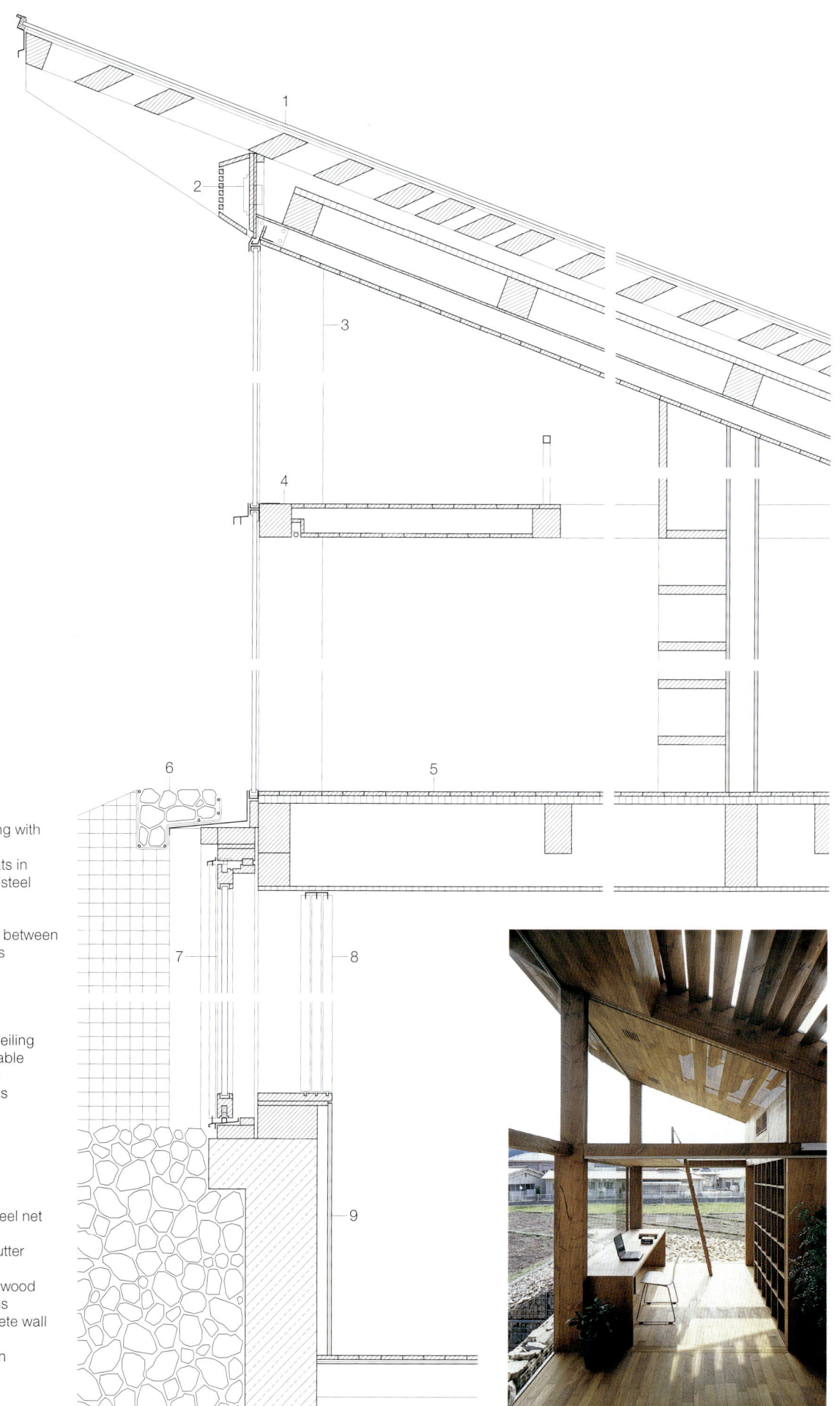

vertical section
scale 1:20

1 roof construction:
 12 mm float-glass covering with sealed butt joints
 60 × 10 mm aluminium flats in 70 × 12 × 2 mm stainless-steel channel sections
 105 mm pine joists
 185 mm ventilation cavity between 120 × 300 mm pine rafters
 24 mm plywood
 120 mm cavity
 6 mm plywood
 75 mm ventilation cavity
 15 mm teak suspended ceiling
2 natural ventilation, adjustable
3 120 × 240 mm pine posts
4 120 × 120 mm pine beams
5 floor construction:
 15 mm teak parquet
 28 mm plywood
 120 × 300 mm pine joists
 15 mm pine-veneered plywood soffit
6 stone rubble in welded-steel net
7 lifting-sliding window
8 three-part teak sliding shutter
9 wall construction:
 15 mm pine-veneered plywood
 40 × 40 mm timber battens
 275 mm reinforced concrete wall
10 three-part sliding window
11 12 mm float-glass partition
12 60 mm teak sliding gate
13 3 mm stainless-steel eaves section

48

Project details:
Usage: private residence
Construction: timber framework
Total internal volume: 722 m³
Total usable area: 271 m²
Site area: 700 m²
Date of construction: 2005
Period of construction: 2004–2005

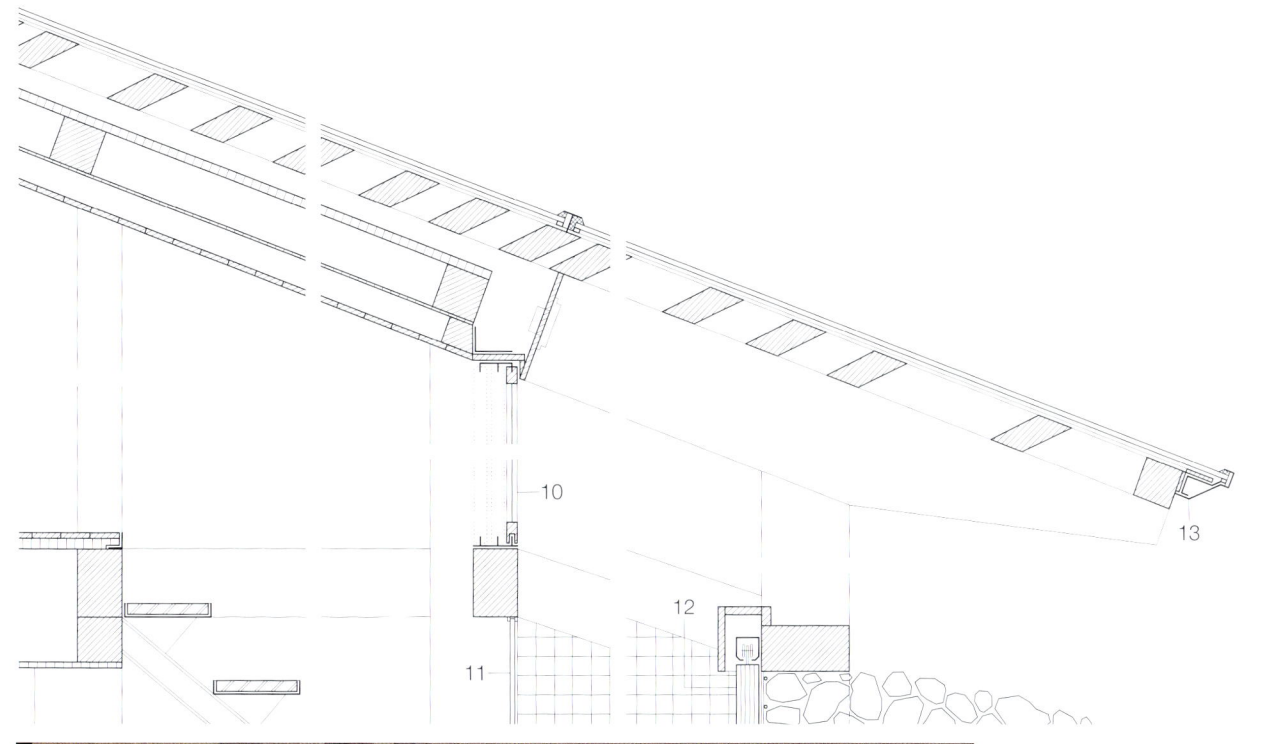

House in Aitrach

Architects: SoHo Architecture, Augsburg

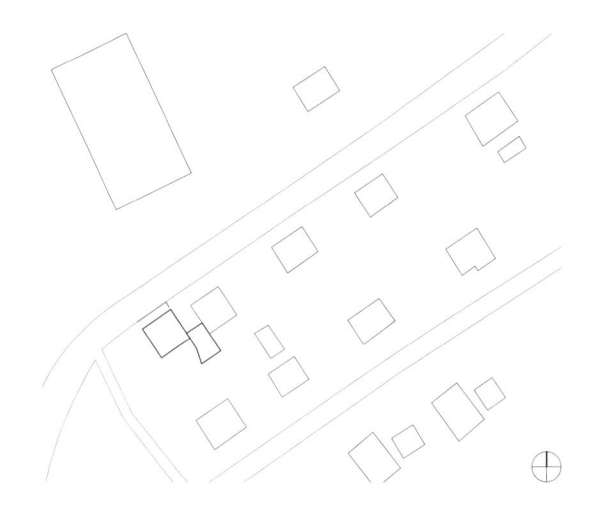

Reference to traditional rural forms with new impulse

In the small-scale village structure of Aitrach in southern Germany, this house, which marks the beginning of the community, stands out as something exceptional. Despite its standard proportions and roof pitch, the outer skin of untreated larch boarding and the closed, black steel fence (which provides sound protection and visual screening) introduce a modern note to the neighbourhood. Even more unusual are the narrow window strip in the street facade and the black painted garage, which contrast greatly with the traditional family home.

Internally, the architects have oriented their design on the traditional dwellings of the area, the almost square-shaped buildings usually being divided into four equal rooms. Here, however, a sense of spatial continuity is created by the more open layout, which can be divided and articulated at will by means of sliding wall elements. This spatial quality is enhanced by the large areas of glazing that continue from one room into another and offer views into a variety of differently planted gardens.

Low-energy house in element construction technique

The low-energy house has a prefabricated timber post-and-beam structure set on a reinforced concrete basement. Although the roof tiling is typical of the area (economical roof tiles were selected) and serves to integrate the building into its environment, there are no eaves projections here. All external walls are moisture-diffusing and have a services layer on the inside. The exposed, stacked-plank intermediate floor is supported by the horizontal rails of the construction and do not pierce the external prefabricated timber walls. An impermeable skin was developed with side and bottom-hung sashes all being the same size, but set in the outer skin at different levels. A "cheap trick" which creates various spatial effects internally: in one case, a high window sill; in another, a high lintel.

The omission of a fragile vapour barrier and the design of simple details enabled the client to execute a large part of the internal fittings himself. In the planning stage it was agreed that the external cladding, the steel fence, the floor finishes and the landscaping would be carried out by the client. The architect took on the role of advising and encouraging the client in matters of quality and design.

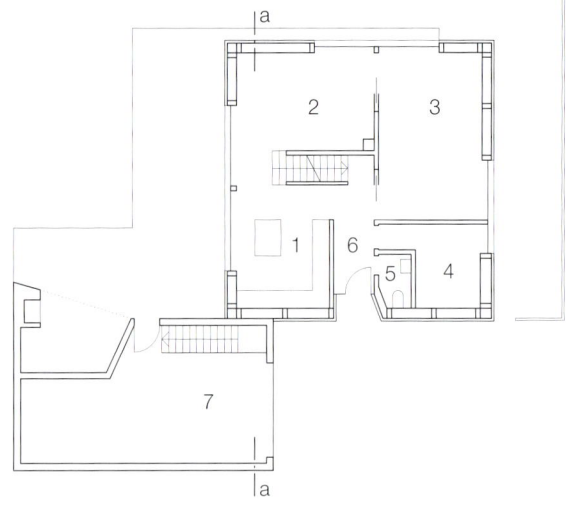

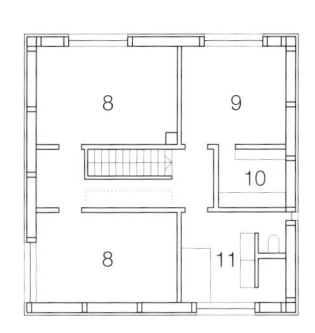

Project details:
Usage: private residence
Construction: timber
Internal ceiling height: 2.6 m ground floor
Total cost of construction: 230,000 Euros
Cost per m² residential area: 1,337 Euros
Total internal volume: 775 m³
Total storey area: 172 m²
Residential area: 132 m²
Built site are: 86 m² + 48 m² (garage)
Site area: 475 m²
Period of construction: 14 months
Date of construction: 2004

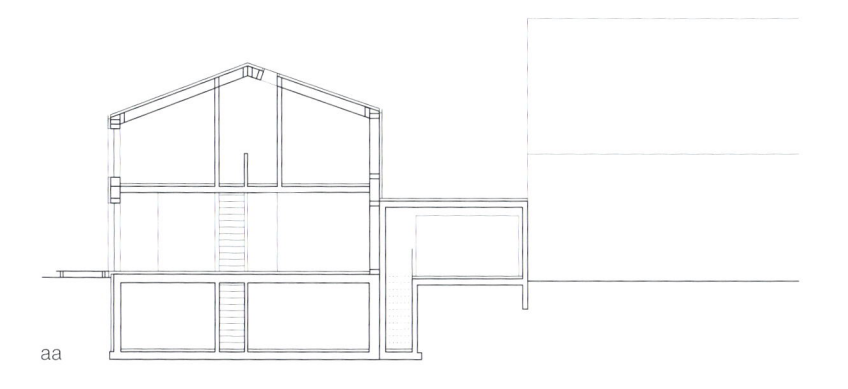

site plan
scale 1:2000
floor plans • section
scale 1:250

1 kitchen
2 dining
3 living
4 utilities
5 WC
6 hall/cloakroom
7 garage
8 child's bedroom
9 bedroom
10 dressing
11 bathroom

aa

vertical sections
scale 1:20

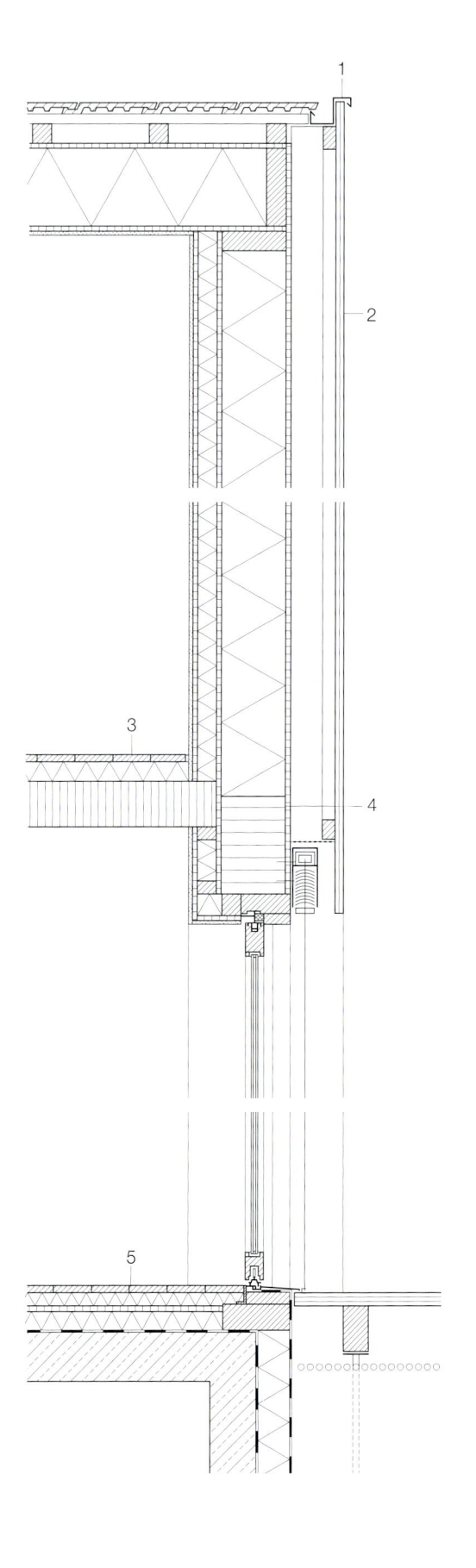

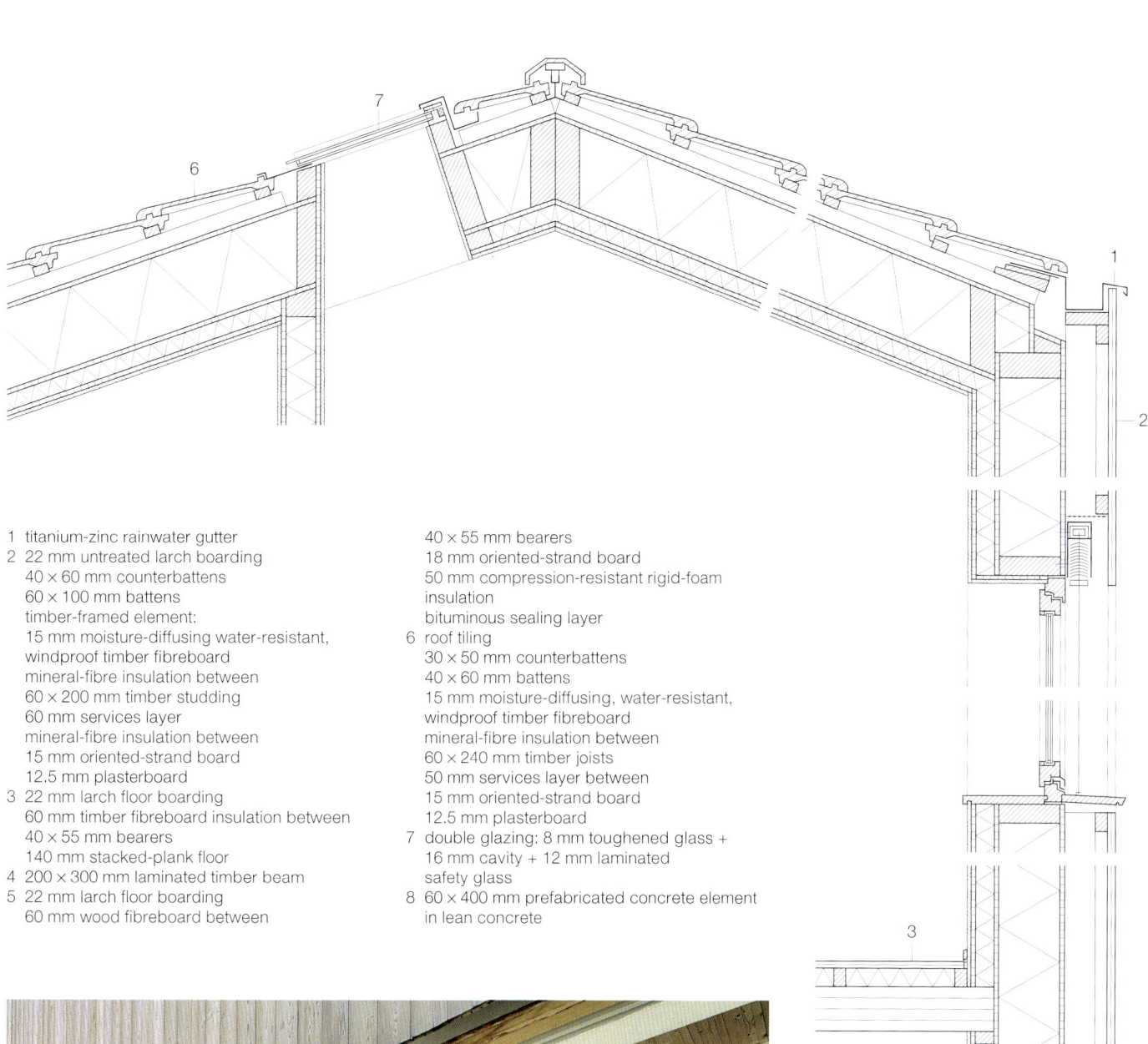

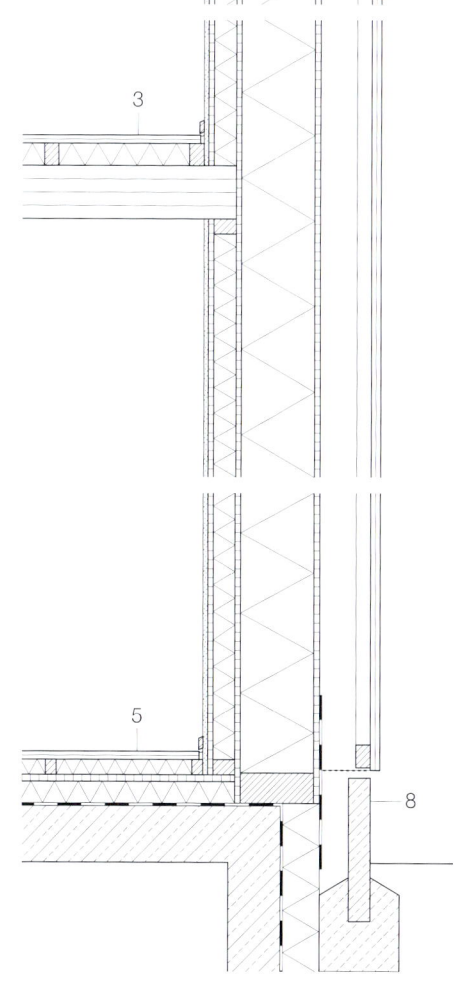

1 titanium-zinc rainwater gutter
2 22 mm untreated larch boarding
　40 × 60 mm counterbattens
　60 × 100 mm battens
　timber-framed element:
　15 mm moisture-diffusing water-resistant,
　windproof timber fibreboard
　mineral-fibre insulation between
　60 × 200 mm timber studding
　60 mm services layer
　mineral-fibre insulation between
　15 mm oriented-strand board
　12.5 mm plasterboard
3 22 mm larch floor boarding
　60 mm timber fibreboard insulation between
　40 × 55 mm bearers
　140 mm stacked-plank floor
4 200 × 300 mm laminated timber beam
5 22 mm larch floor boarding
　60 mm wood fibreboard between
　40 × 55 mm bearers
　18 mm oriented-strand board
　50 mm compression-resistant rigid-foam
　insulation
　bituminous sealing layer
6 roof tiling
　30 × 50 mm counterbattens
　40 × 60 mm battens
　15 mm moisture-diffusing, water-resistant,
　windproof timber fibreboard
　mineral-fibre insulation between
　60 × 240 mm timber joists
　50 mm services layer between
　15 mm oriented-strand board
　12.5 mm plasterboard
7 double glazing: 8 mm toughened glass +
　16 mm cavity + 12 mm laminated
　safety glass
8 60 × 400 mm prefabricated concrete element
　in lean concrete

53

Weekend House at St Andrews Beach

Architects: Sean Godsell Architects, Melbourne

Reduced construction and maintenance costs
Building envelope of pre-oxidised grates
Minimal comfort

St. Andrew's Beach on Victoria's Mornington Peninsula is one of the few places in Australia where you can build directly on the coast. This elevated weekend house enjoys superb views over the ocean and harmonises in both colour and form with the surrounding landscape. This design is, however, a new interpretation of the Australian weekend house. The transparent building skin, made of pre-oxidised steel grates, serves only to provide shade for the large continuous timber deck. Mounted upon this deck are two separate units – one containing the living-dining area, while the other unit accommodates three introspective bedrooms and a study. When moving from one space to another, the occupants are exposed to the natural elements and, particularly in summer, to extreme temperatures. The clients wanted a retreat where they could get back in touch with nature; with the wind and the weather, after a week in the city working in air-conditioned offices at constantly controlled temperatures of 22 °C. The steel skeleton is slung between two pairs of columns; room-high diagonal struts provide bracing in the longitudinal direction. Alone the living area benefits from transparent thermal insulation, fitted below the corrugated GRP roof, which allows diffuse light to permeate into the interior. On the western side glass doors open up the living area to the veranda, while the bedrooms are darker and more enclosed. The rust-brown of the oxidised steel and the timber flooring create a unified, homogeneous appearance.

Due to the simple, yet well thought out, detailing and the lack of continuous thermal insulation throughout the building, it was possible to save greatly on construction costs. The facade elements; standard industrial steel grates, were greatly enhanced by their transformation into a residential cladding material. Additionally, the usage of directly recycled construction elements and the creation of a low maintenance external envelope enabled substantial cost savings to be made.

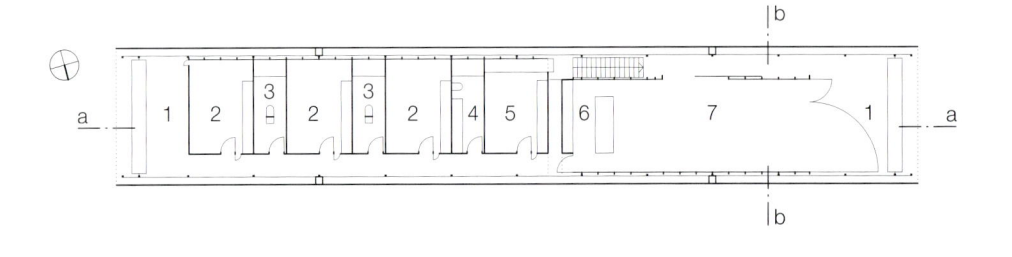

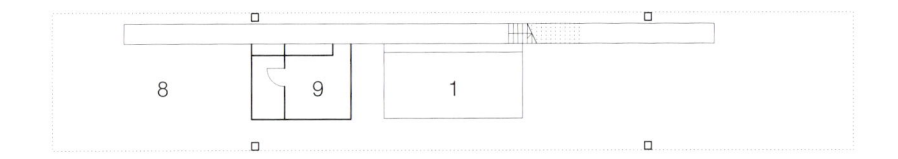

section · floor plans
scale 1:400

1 deck
2 bedroom
3 bathroom
4 bathroom/laundry
5 workroom
6 kitchen/dining
7 living
8 carport
9 storage

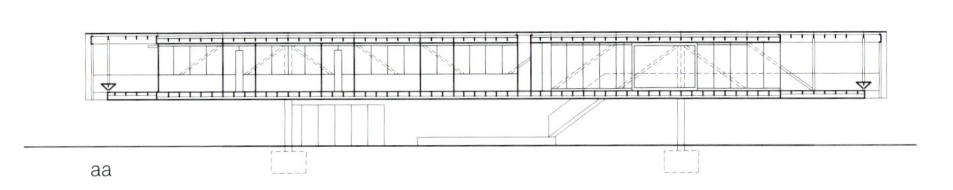

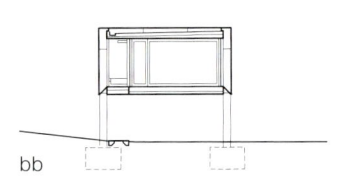

55

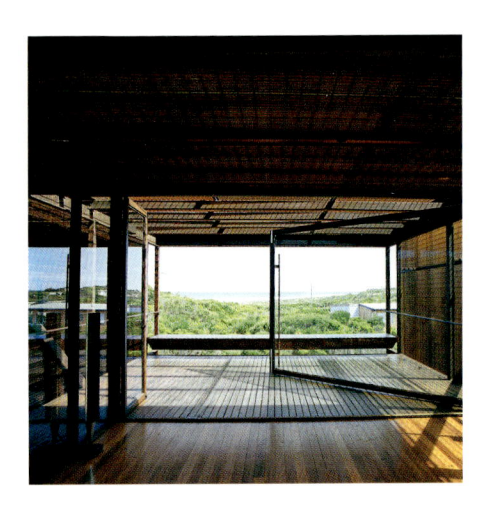

cross section
longitudinal section
scale 1:20

1 30 × 3 mm pre-oxidized steel grating
2 point fixing:
   Ø 150 mm pre-oxidized steel disc
   Ø 12 mm galvanized steel rod
   40/40 mm steel angle
3 double layer GRP
4 thermal insulation
   10 mm fibreglass honeycomb thermal insulation
5 390 × 100 mm pre-oxidized steel channel section
6 150 × 100 mm steel RHS girder
7 light fitting; 1.6 mm folded steel sheet
8 50 × 50 mm pre-oxidized steel SHS
9 50 × 50 mm pre-oxidized steel T-section bracket
10 100 × 100 mm pre-oxidized steel channel section strut
11 frame for sliding doors;
   50 × 35 mm pre-oxidized steel RHS
   8.4 mm toughened glass, low-e coating
12 handrail; Ø 32 mm pre-oxidized steel tube
13 bench; 4 mm steel sheet folded around 150 × 100 mm steel RHS
14 19 mm recycled Chinese tallow boards
   150 × 50 mm timber beam
15 floor construction;
   19 mm recycled Chinese tallow boards
   40 mm composite timber board
   60 mm polystyrene thermal insulation
   240 mm lattice girder with timber chords and metal struts
16 10 mm folded pre-oxidized steel sheet
17 Ø 18 mm steel rod, e = 1800 mm
18 400 × 400 mm pre-oxidized steel SHS column
19 75 mm reinforced concrete pavers

Project details:
Usage: weekend house
Construction: steel skeleton
Internal ceiling height: 2.6 m
Total cost of construction: no data
Cost per m² usable area: no data
Total internal volume: 800 m³
Total built area: 280 m²
Total usable area: 260 m²
Total site area: 3,900 m²
Construction period: 12 months
Date of construction: 2005

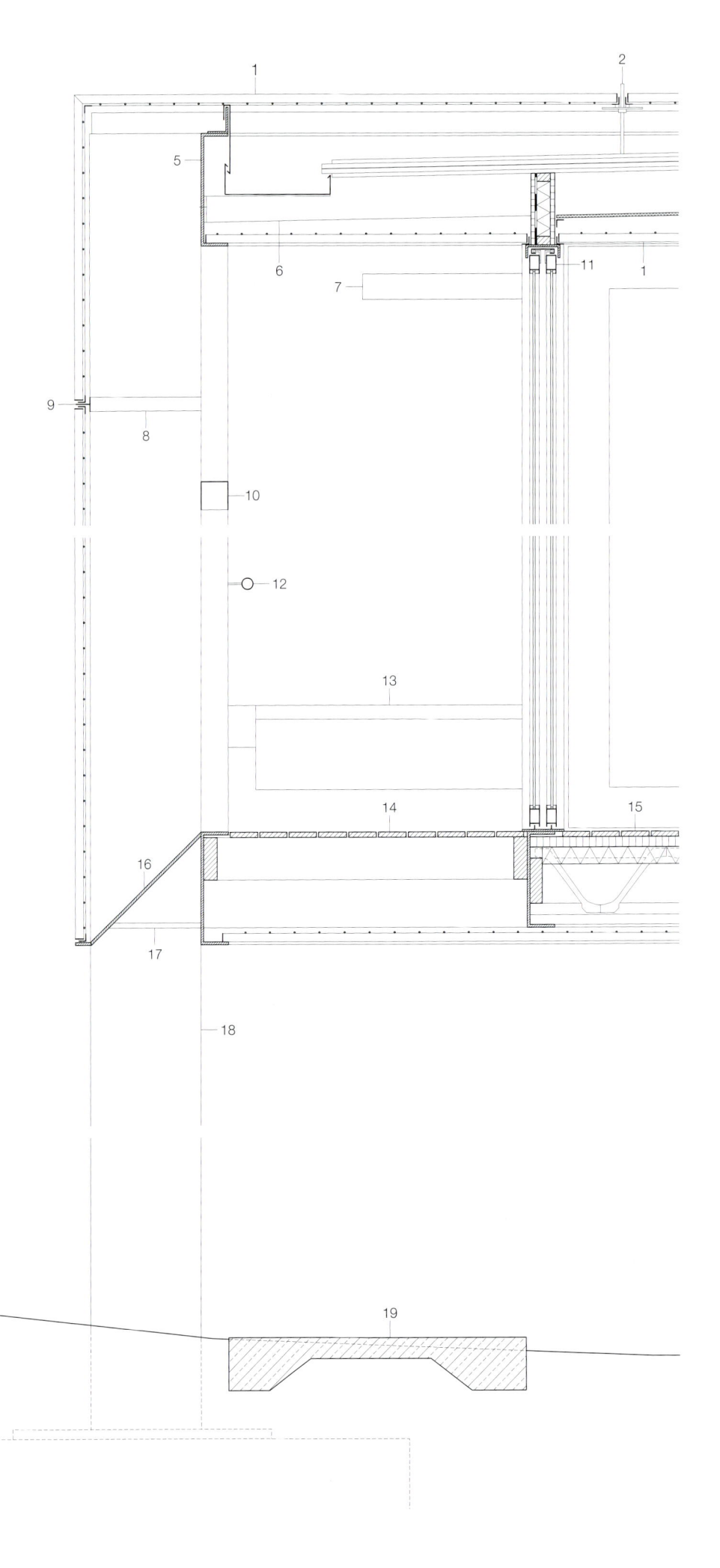

57

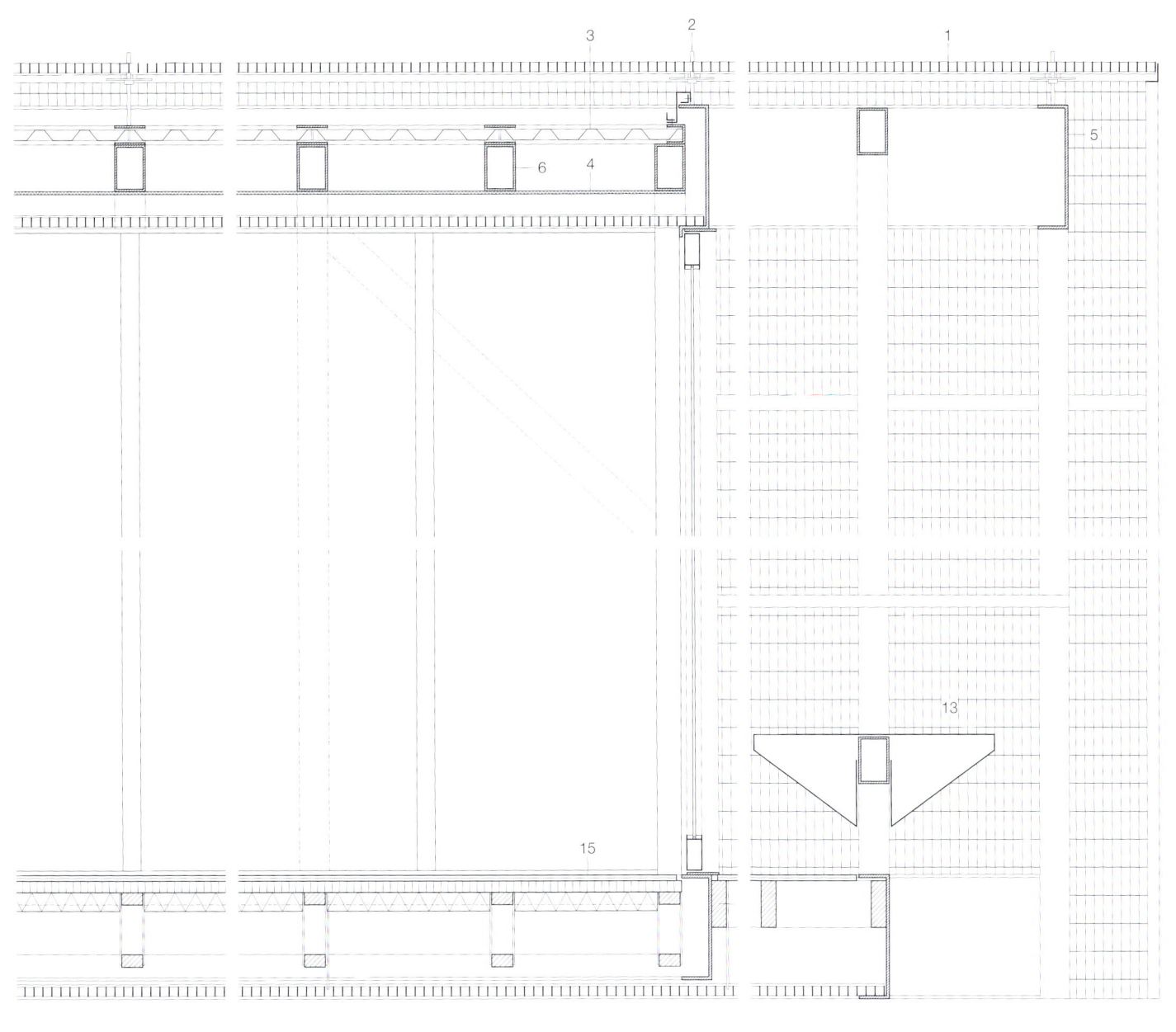

House on Lake Laka

Architect: Peter Kuczia, Osnabruck

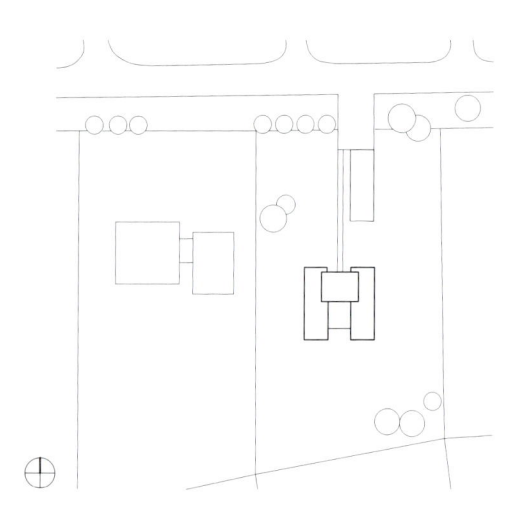

Single family house with solar technology
Reduced construction and lifecycle costs
Simple, well thought-out details
Usage of local materials

Like a chameleon, this house blends with its surroundings in Pszczyna, in the region of Upper Silesia. Colourful panels within the timber facade reflect the tones of the landscape and the window reveals, clad in fibre cement, frame images of the countryside.

Sustainable energy concept

The built form is designed to optimise the absorbance of solar energy, with approximately 80% of the building envelope facing south.
Although the building is symmetrical, the internal zones are arranged asymmetrically according to function. The single storey, quadratic living space on the ground floor is externally clad with untreated Larch boarding. Additional solar energy is gained by the set-in glazed patio. Solar collection panels, for the heating of water, are located on the roof surface which is pitched at the optimum angle for solar energy absorption. Long-term reductions in energy consumption and carbon-dioxide discharge are also achievable with a photo-voltaic system planned for the future.
Located in the so-called "black box", a three storey structure clad with dark grey fibre cement panels, are a work studio with gallery, and a guest room which benefits from panoramic views of the adjacent reservoir through picture windows. The dark facade is warmed by the sun, reducing heat loss to the environment. The passive and active solar energy concepts and a high standard of thermal insulation are further enhanced by a ventilation plant with thermal recovery system.

Cost saving strategies

The design of the project was determined by the twin goals of low lifecycle costs and a reduction in construction costs. Due to the simple, but well thought-out, details the construction of the house did not cost more than a conventional house in Poland.
Noticeable savings were made by the application of traditional building techniques and the use of local materials and directly recycled building elements. The facade sub-construction is based upon the use of a single layer of slanting battens rather than battens and counter battens, thereby reducing material costs. All floor finishes are of sanded and polished concrete, which is considerably more economical in Poland than conventional multi-layered flooring.

Project details:
Usage: single-family house
Construction: brickwork, timber
Total construction cost: 95,000 Euros
Cost /m^2: 422 Euros
Total internal volume: 730 m^3
Total built area: 226 m^2
Usable area: 175 m^2
Internal ceiling height: 2.23–4.20 m
Site area: 1,990 m^2
Built site area: 165 m^2
Date of construction: 2007

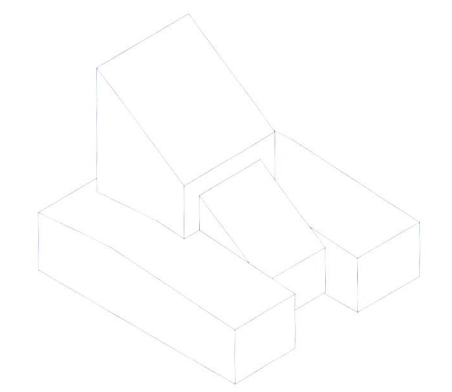

site plan
scale 1:1500

section · floor plans
scale 1:250

1 living
2 dining
3 kitchen
4 store room
5 porch
6 cloakroom
7 laundry
8 bathroom
9 bedroom
10 terrace
11 studio
12 gallery
13 atrium void
14 guest room
15 corridor

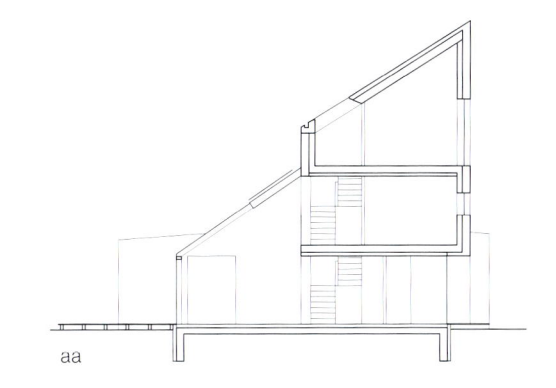

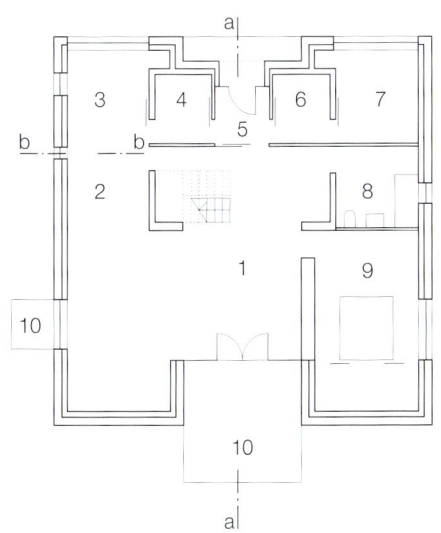

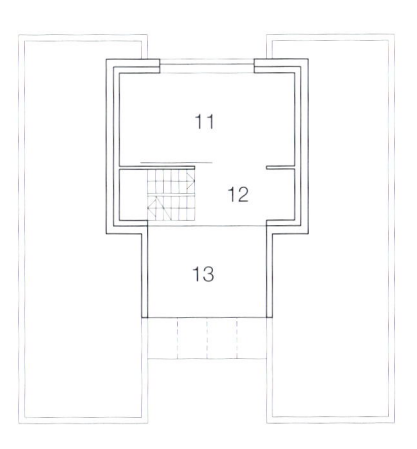

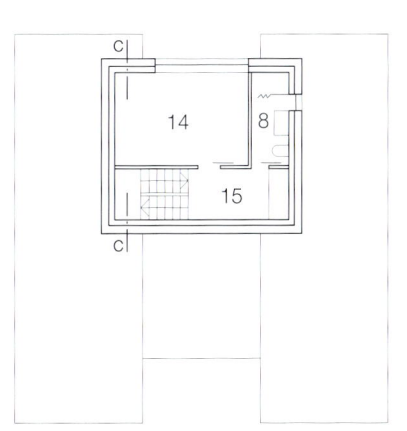

aa

vertical sections
horizontal section
scale 1:20

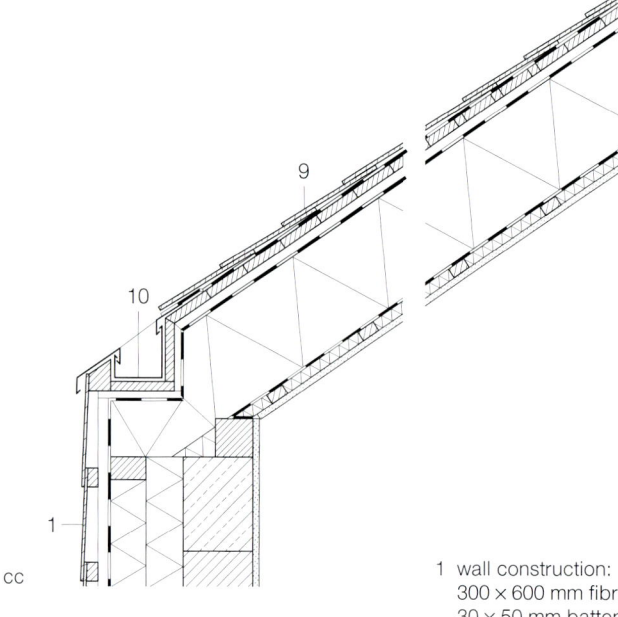

1 wall construction:
  300 × 600 mm fibre cement panels
  30 × 50 mm battens,
  30 × 50 mm counterbattens
  30 mm ventilation cavity
  diffusion permeable membrane
  double layer 200 mm mineral wool
  thermal insulation
  between 100 × 60 mm timber framing
  190 mm perforated brickwork
2 metal flashing
3 flat roof construction:
  60–80 mm extensive vegetation
  bituminous roofing with root protection layer
  24 mm timber cladding
  30 mm ventilation cavity
  diffusion permeable membrane
  240 mm mineral wool thermal insulation
  vapour barrier
  40 mm thermal insulation
  15 mm plasterboard
4 timber window with double glazing
5 window reveal: 12 mm fibre cement panels
6 external wall construction:
  21 × 150 mm larch boarding
  30 × 50 mm battens, slanting
  30 mm ventilation cavity
  diffusion permeable membrane
  double layer 200 mm mineral wool
  thermal insulation
  between 100 × 60 mm timber framing
  190 mm perforated brickwork
  15 mm plasterboard
7 160 mm plinth insulation
8 floor construction:
  80 mm concrete screed, sanded and
  polished
  20 mm impact-sound insulation,
  sealing layer, 120 mm reinforced concrete
  120 mm perimeter insulation, membrane
9 pitched roof construction:
  300 × 600 mm fibre cement roof panels
  roof paper, 24 mm timber cladding
  30 mm ventilation cavity
  diffusion permeable membrane
  300 mm mineral wool thermal insulation
  vapour barrier
  30 mm insulation
  15 mm plasterboard
10 zinc metal box gutter

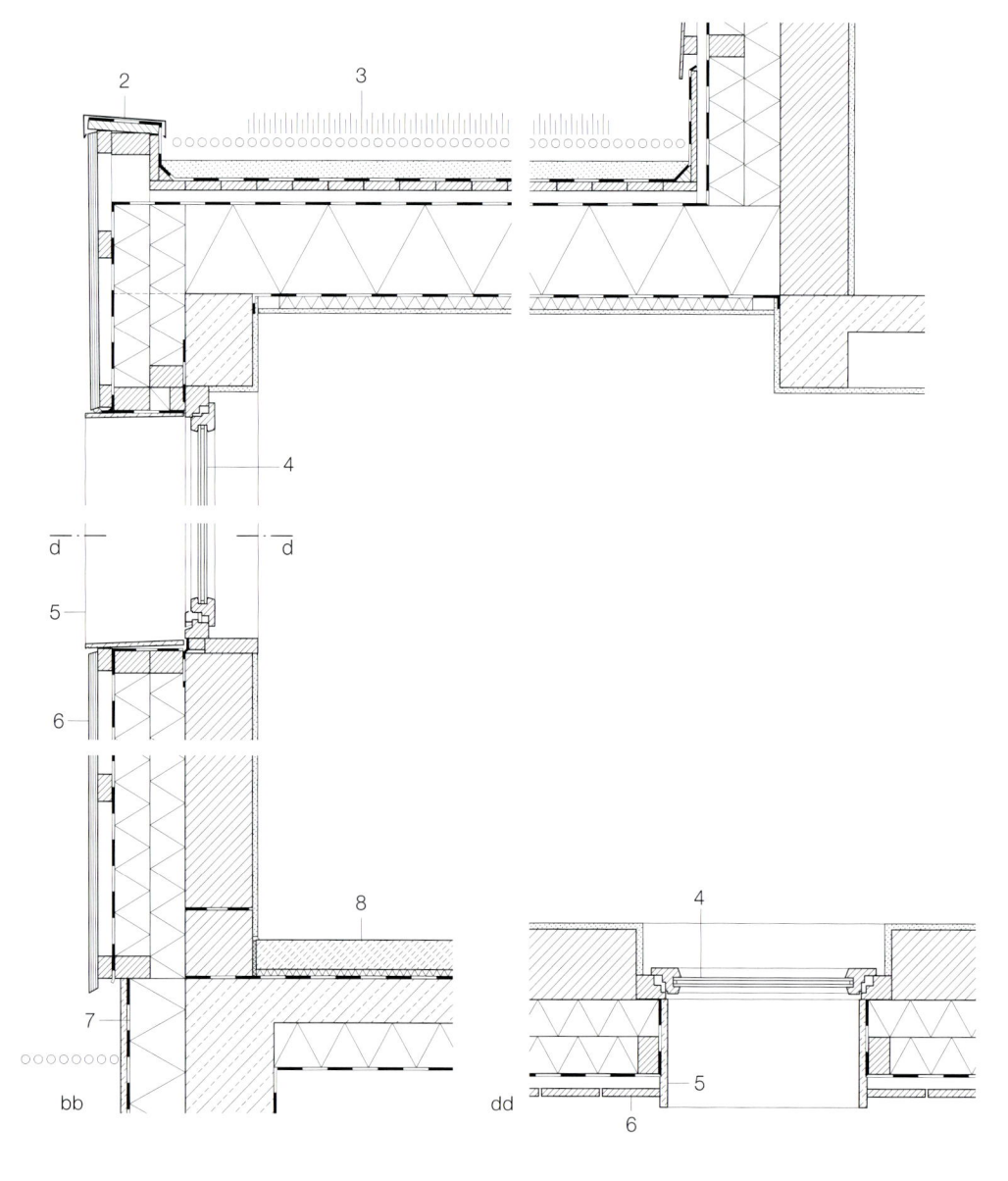

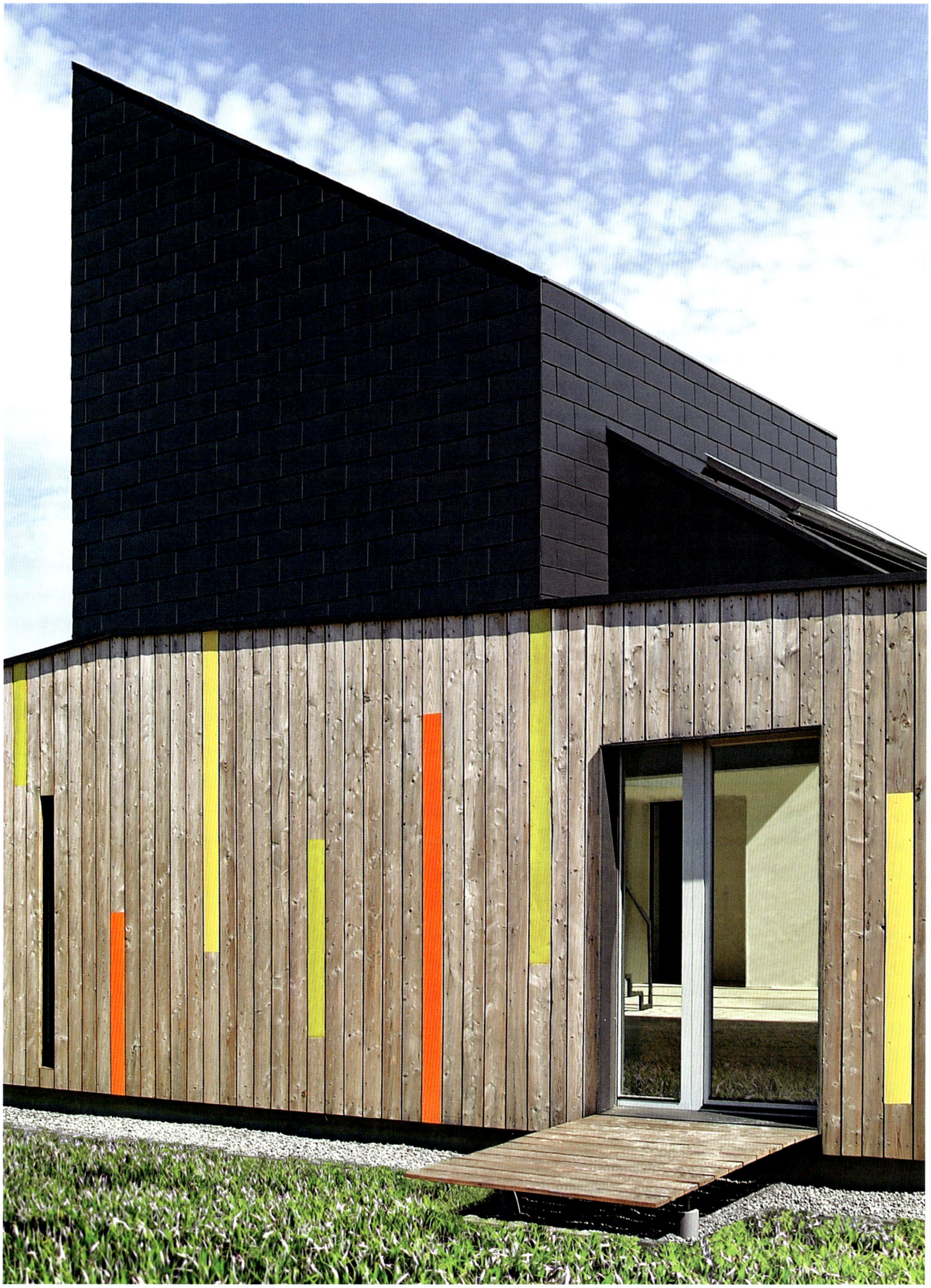

Housing Development in Neu-Ulm

Architect: G. A. S.-Sahner, Stuttgart

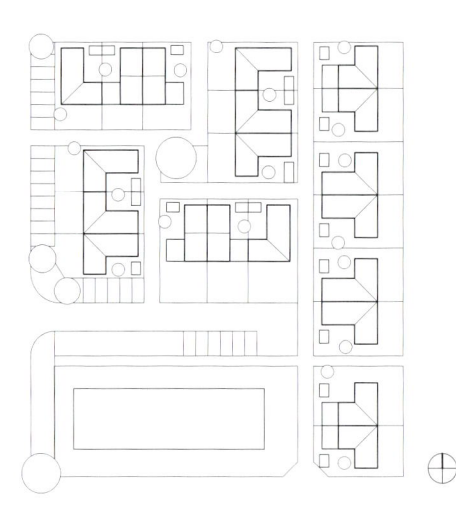

Economical and ecological construction system
Small component structures

South of Neu-Ulm one comes across extensive areas of land that were formerly occupied by military barracks. The existing structures, whether deserted, forgotten military buildings or new constructions, appear to have been placed randomly within this expanse of fallow land without any thought to urban layout. This modest residential development of twenty houses is situated on the edge of the former military area adjoining agricultural land. They are the result of a building society competition for the design and development of a low-cost, environmentally friendly house construction system. The urban development claims a high density – a floor-space index of 0.8 – yet conveys a sense of human scale reminiscent of village structures with central, open "green" and interconnecting lanes. The development plan was drawn up by the architect and developer in collaboration with local authorities, and received specific exemption in the overlapping of setback areas.

Individual house combinations

The small entrance courtyards to the L-shaped houses and the associated garden sheds provide an important element of privacy for the residents. At first glance it is not evident that all the houses have an identical ground-floor layout and are developed from the same modular construction system. With their various roof forms and the scope they provide for different groupings, the houses create an interesting and polymorphic ensemble.

The fundamental principle behind the development is, of course, the basic construction module of the individual houses which is capable of providing floor layouts ranging from 123 to 143 m^2. The minimum volume comprises two rooms and a staircase, with a kitchen, bathroom and WC linked to a central services core. This nucleus can be extended by adding further individual rooms. The maximum volume is a seven-room dwelling extending over three floors. The rooms are similar in size and form, so that the houses can accommodate a wide range of user requirements. If required, a self-contained ground-floor apartment can also be divided off at a later date without major constructional intervention. Other variations are possible with the addition of basement spaces and the selection of different roof elements.

The construction modules themselves are predominantly prefabricated and consist of a fixed range of components, for example a pre-selected window schedule. This modular construction system of large-format, dimensionally precise wall and ceiling elements makes simple connections and a chiefly dry form of construction possible, reducing construction time on site. The exterior carcass of a house can be completed on site within two days and the flexibility of the concept allows various local building materials to be used, as available.

Project details:
Usage: 20 semi-detached houses
Construction: brickwork
Internal ceiling height: 2.5–4.1 m
Cost of construction: 1.82 million Euros
Cost per residential area: 817–941 Euros
Total floor area: 9,640 m^2
  123–143 m^2
Total site area: 3,879 m^2
Energy requirement for heating: 60.83–68.00 kWh/m^{2}a
Date of completion: 2000

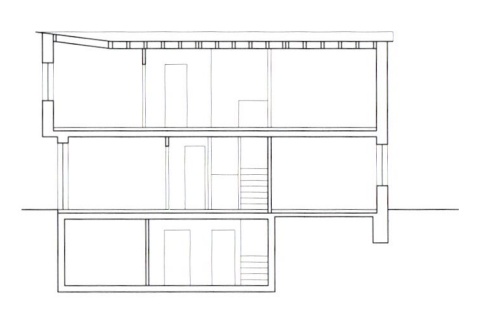

aa

site plan   scale 1:1500
house type CX 110: floor plans • section
scale 1:250

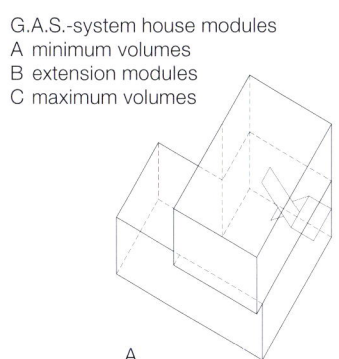

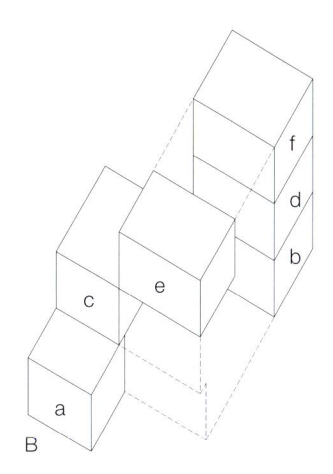

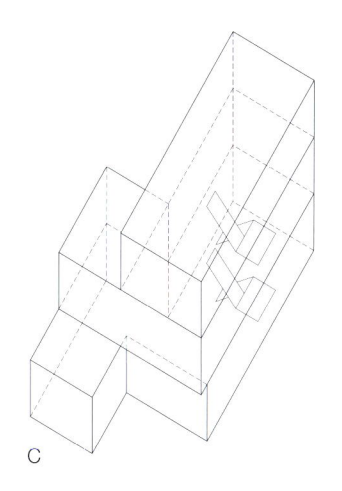

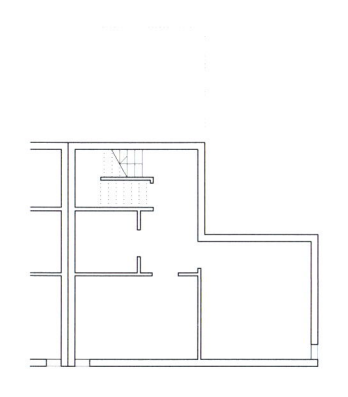

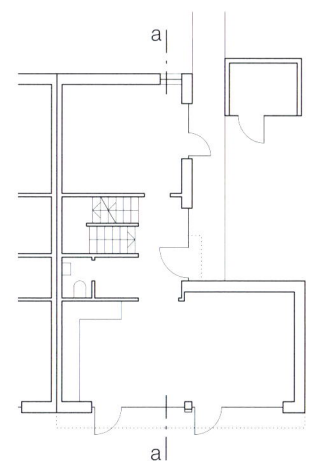

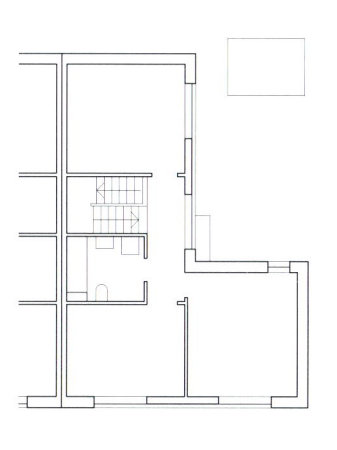

G.A.S.-system house modules
A minimum volumes
B extension modules
C maximum volumes

A

B

C

a
b
c
d
e
f

63

house type BX 90: section • floor plans
scale 1:250

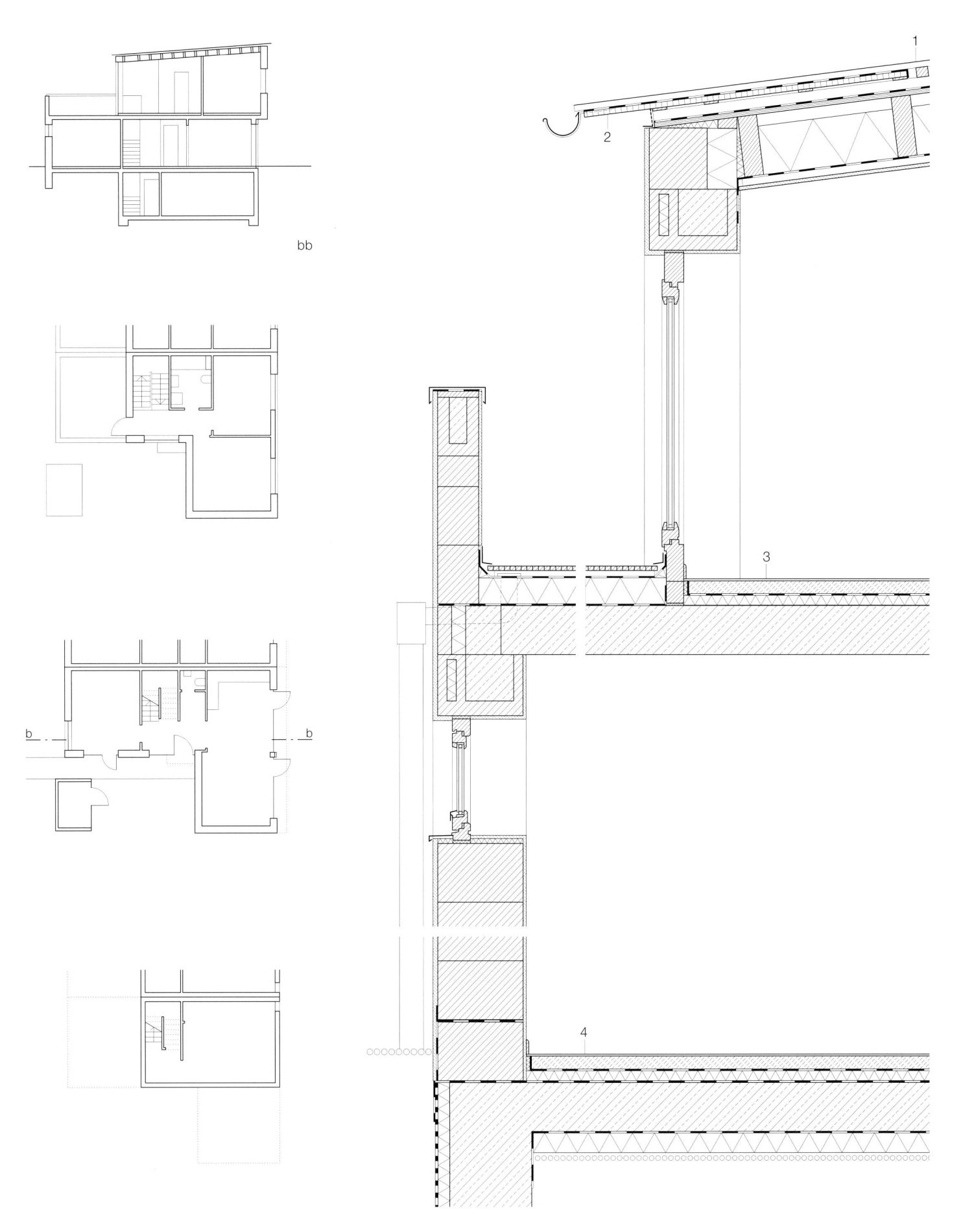

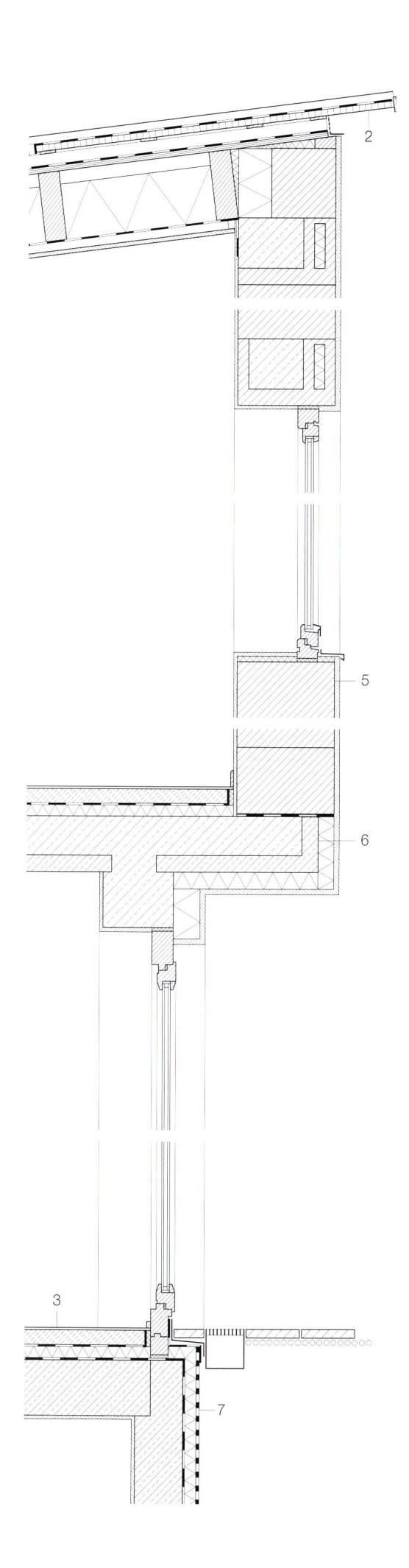

vertical section
scale 1:20

1 roof construction:
   18 × 76 mm corrugated sheet aluminium
   50 × 40 mm battens, counter-battens
   waterproof membrane
   24 mm sawn softwood boarding
   200 mm mineral-fibre insulation between
   80 × 220 mm softwood rafters
   vapour barrier
   48 × 28 mm softwood battens
   12.5 mm plasterboard
2 25 mm three-ply laminated softwood panel
3 first floor construction:
   carpeting or PVC
   50 mm screed
   polyethylene separating layer
   50 mm thermal/impact-sound insulation
   200 mm prefabricated concrete element
4 ground floor construction (no basement):
   carpeting or PVC
   50 mm screed
   polyethylene separating layer
   50 mm thermal/impact-sound insulation
   waterproof membrane
   200 mm in-situ concrete slab
   polyethylene separating layer
   80 mm perimeter insulation
5 365 mm rubbed-brick wall
   (λ_R = 0.11 W/mK)
6 prefabricated concrete element as permanent formwork with 60 mm thermal insulation
7 external basement wall construction:
   textured drainage sheet
   50 mm perimeter insulation
   waterproof sealing layer
   180 mm prefabricated concrete elements
   10 mm render

65

Different Forms of Construction for a Modular Unit Building System

Georg Sahner

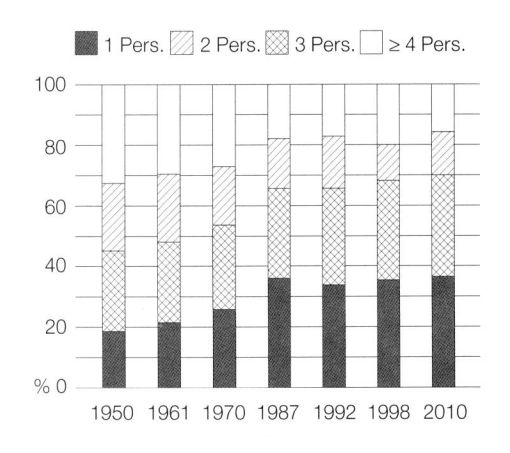

Table 1:
until 1987: Federal Republic of Germany (old)
since 1992: Federal Republic of Germany (new)

The idea underlying the development of a programme of system housing was based on social developments in Germany in recent years. The driving force behind those architects who attempt to standardise housing construction and rationalise manufacture today, is no longer simply the desire to optimise manufacture and costs. Architects like Charles Eames, Jean Prouvé, Richard Buckminster Fuller and Konrad Wachsmann, who helped shape the history of industrialised building, all attempted to combine the solution of social problems with an economic use of resources. Today, the creation of living spaces appropriate to contemporary lifestyles is more important. Even when developing a cost efficient house of prefabricated elements and modules it is possible to meet the demands of layout options, high level of technical quality and product diversification without incurring additional costs. Compared with figures for earlier decades, predictions for the size of households in the year 2010 (see table 1) show a significant shift towards smaller units, with a large proportion of single person and single-parent households. In the majority of German cities 42 % of all children reside with one parent and one and two person households dominate over households with children. Families with three or more children are, statistically speaking, irrelevant. Household sizes both increase and decrease. In Germany, people become homeowners relatively late in life (at an average age of 38, compared with 26 in the Netherlands). Among the reasons for this are the high costs of construction and land. Whoever wants to accrue real estate must be prepared not only to save for a long period of time, but also to take on a larger and longer lasting credit. The development of flexible forms of construction that can be extended or reduced in size to take account of changing family structures and that can be acquired at a modest initial price, which means a smaller risk for the household, is a central aspect of our work.

Modular concept

The G.A.S. System House is based on a modular, additive principle. A limited number of building units (house modules) are so combined as to produce different complete houses. House series are thereby produced from repetitive house modules. The system comprises three series of modules: basic modules (x, y and z), extension modules (a, b, c, d and e – see p. 63) and supplementary modules. The basic modules contain the essential components of a house:
- entrance and staircase
- services duct,
- WC, bathroom, etc.

The extension modules are pure room modules, while the supplementary modules consist of roof, basement, conservatory elements and the like. A typical house with 74 m² residential floor area (II. BV), for example, would comprise two basic modules and two supplementary modules, including at least one roof module. The combination of these modules allows the creation of various housing types. The intention is to achieve combination alternatives in the future at no extra cost. At present the development of the house series has been stopped, therefore only houses with the areas of 70, 90, 110 and 130 m² are available as:
- detached houses
- semi-detached houses
- terraced houses
- chain houses
- stacked houses (up to six storeys)
- courtyard houses
- houses for hillside sites

The modules are designed to allow a free internal layout with scope for subsequent modifications. This economical and highly alterable system can react to changes and new demands from the housing market. At present, using only 16 modular elements, it is possible to build 18 different dwelling types. In combination with the eight supplementary modules, the basement and roof modules, this allows 118 housing forms to be created, not to mention the internal variations that are possible. Experience shows that great scope for variation, in addition to the various construction systems which will be discussed below, requires intelligent project management but also incurs high costs at the preliminary planning stage.

Prefabrication

The fundamental reason behind subdividing the modules up into individual construction elements is the reduction of the exorbitant transport costs associated with prefabricated elements. In addition to which, the regional conditions and cost advantages could be considered and various construction systems could be developed:
- prefabricated brick wall elements with pre-cast concrete floors and stairs,
- prefabricated aerated concrete panels with aerated concrete floor elements,
- expanded-clay-concrete wall panels with solid concrete floors,
- large format sand-lime brick elements with filigree floor elements,
- timber stud elements, complete with finishes and fittings.

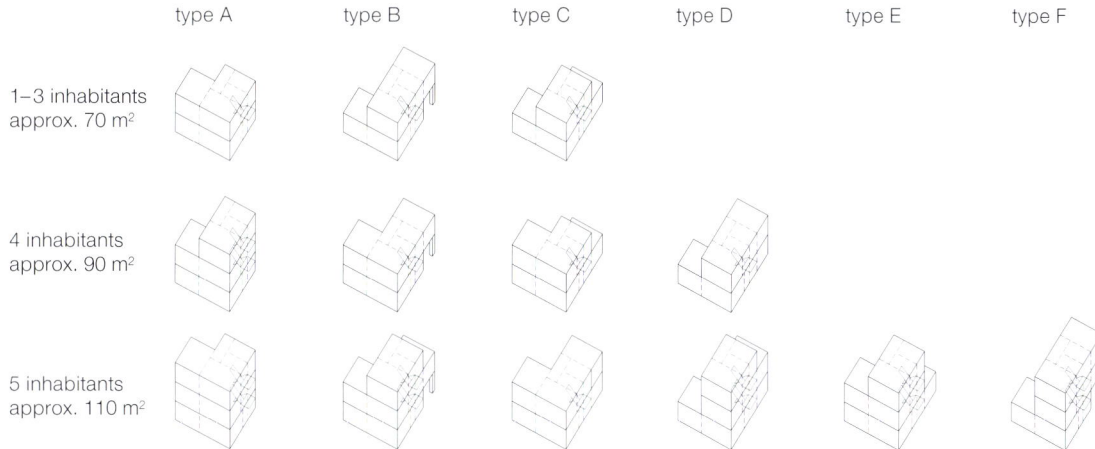

	type A	type B	type C	type D	type E	type F
1–3 inhabitants approx. 70 m²						
4 inhabitants approx. 90 m²						
5 inhabitants approx. 110 m²						

G.A.S. System House: Different configurations achieved with various combinations of modules

The detailing of the elements was based on the following criteria:
- rapid assembly process, optimisation of connections, simplicity,
- dry form of construction, highest possible standards, including render when possible,
- repetitive units, application of standard elements
- guaranteed quality through precision work.

It has been proven on site that exceptionally efficient time management is fundamental to the success of an assembly building site, only then can costs be saved. Further, the individual transport runs must be fully optimised in order to ensure better economy. Element sizes and type were often determined by transport alternatives.

Prefabricated brick construction

In contrast to timber construction, it has not proved possible with brick construction to achieve cost savings through the use of modern working techniques and organization. High labour costs normally lead to the development of industrialized manufacturing methods; but these require large investments, and in Germany, at least, the building industry has preferred to employ cheap labour rather than abandon traditional brick construction techniques. This has led to problems in maintaining quality standards of external brickwork. Ensuring quality and implementing a quick, dry construction process were two of the main objectives in developing the system house. A special block with a high degree of thermal insulation was designed expressly for this purpose. Due to the density of 0.6 kg/dm² and geometry of the brick, a thermal conductivity of only 0.11 W/mK was achieved. The wall elements are manufactured to a thickness of 36.5 cm and, with rendering and internal plaster, achieved a U-value of 0.28 W/m²K (without windows). The maximum panel sizes are 2.80 × 14.00 m. The wall construction meets the new energy standards of 2007 without requiring additional thermal insulation. The latest developments in the field of brickwork attain levels of 0.08 W/mK, and even 0.06 W/mK is expected in the near future. Low energy standards will soon be able to be met by solid wall elements without the need for additional composite thermal insulation systems.

The maximum assembly time for the carcass structure of a house in prefabricated brick unit construction is two days (including the floor elements and roof construction). The dimensional precision of the computer-controlled brickwork is so high that the internal plaster can be executed simply as a smooth stopping layer. The large numbers of completed brick system houses indicated the wide acceptance of this form of construction by the clients. Construction costs remain within acceptable ranges when the logistics of the production processes are well organised.

Large format sand-lime brick elements and aerated concrete wall elements

Brickwork elements can reach their limitations when particularly high levels of noise protection are necessary. System houses have also been erected in numbered, pre-cut sand-lime building slabs. With a wall thickness of 24 cm plus the necessary insulation, these provide the same results as the brick type, particularly in the area of noise protection. Aerated-concrete wall elements with a thickness of at least 20 cm can be just as quickly and economically erected as brick elements, but with a thermal conductivity of 0.14 W/mK they require more elaborate thermal insulation.

Expanded-clay expanded-glass wall panels

In the system house type "Living x" wall elements in the form of a 17.5 cm skin of expanded-clay concrete with a 12.5 cm layer of expanded cast-glass have been used. All window elements and installations were integrated in the factory. Although produced in a fully automated process, these elements are still considerably more expensive than comparable brick products. With expanded-clay-concrete units, however, savings may be expected as a result of their quick assembly and the fact that no plaster is required internally. This is particularly noticeable in the areas of site programming and the speed with which subsequent works can be commenced.

		Costs timber house (timber stud construction)		Costs massive house (brickwork construction)	
		€	% of total costs	€	% of total costs
300.1	Carcass construction	9379.–	9.5	36166.–	38.2
300.2	Carpentry	44449.–	45.3	5097.–	5.4
300.3	Plumbing/Roof	4003.–	4.1	5335.–	5.6
300.4	Windows incl. glazing	6109.–	6.2	5335.–	5.6
300.5	Entry doors	1714.–	1.7	1895.–	2.0
300.6	Cabinet making incl. doors	1875.–	1.9	2619.–	2.8
300.7	External render incl. scaffolding	5612.–	5.7	10284.–	11.9
300.8	Internal render/Plasterboard	included in 300.2	–	4477.–	4.7
300.9	Metal fitter	455.–	0.5	685.–	0.7
300.10	Screed work	2458.–	2.5	2114.–	2.2
300.11	Tiling	992.–	1.0	713.–	0.8
300.12	Painting	1653.–	1.7	1902.–	2.0
400.1	Heating	8759.–	8.9	8156.–	8.6
400.2	Sanitary installations	5287.–	5.4	5089.–	5.4
400.3	Electrical	3447.–	3.5	2731.–	2.9
300.13	End cleaning	402.–	0.4	441.–	0.5
300.14	Basement room	1631.–	1.7	1631.–	1.7
Total (positions 300 + 400)		98225.–	100.0	94670.–	100.0

Table 2: Total construction costs of different trades for the G.A.S. System House type CX 110: low-energy house with single pitch roof without basement

Timber stud construction

Although prefabricated timber houses are manufactured in a fully automated process, the prices of buildings in this form of construction are not much lower than those for other systems. One reason for this is that the savings from standardisation are usually offset by the elaborate alternative forms and finishes that are offered.

The benefits of serial production could therefore not be fully reaped. The finances necessary for the highly complex production workshops caused the prices of the individual houses to skyrocket. A large German manufacturer in fact applied additional charges of 26 %. In view of this, care was taken in designing the timber system house that individualisation of the dwellings does not undermine the series concept. All construction elements and the external walls remain serial elements. The increased costs of the timber stud construction houses were balanced out against those of the massive construction houses. The great advantage of timber forms of construction lies in the fact that the wall elements are supplied in a largely finished state, which makes a construction period of roughly 4–6 weeks for a house feasible. Another positive factor is the good standard of insulation (16 cm in construction plus additional insulation in the installation level plus additional insulation in the external wall), which indicates positive opportunities in the future. The passive house standard of a heat consumption level of 15 kWh/m²a is more economically and simply achieved in timber than in comparable massive constructions. Other advantages, like installation of heating elements and air-tight connection of windows and doors, offer further opportunity for optimisation in timber system houses.

In conclusion, one can say that the choice of materials for the outer wall elements is decisive in the manufacturing costs of a building. With thermally optimised external walls, it plays an even more decisive role. Higher costs are often incurred with timber-framed or stud buildings as a result of the elaborate external wall construction, which may consist of 10 or more layers, compared to the five or six layers of monolithic masonry walls. Essentially, it can be said that the layered construction of the external walls, in addition to the greatly over-dimensioned timber sections and complicated connec-

Table 3: Building costs for houses with different forms of external wall construction.
Example: G.A.S. System House type CX 110. Residential area: 103.8–107.8 m² (depending on form of construction) acc. to II. BV.

	k-value (wall) [W/m²K]	Carcass + carpentry [€] [€/m²]	House costs [€] [€/m²]	% of total construction costs (cost group 300)	% of total technical services costs (cost group 400)	Period of construction (weeks)	Level of prefabrication	Level of precision
1. Solid, monolithic d = 36.5 cm element brick with λ = 0.11	0.28	41454.– 399.–	93450.– 900.–	82.8	17.2	16	middle	middle
2. Massive double-leaf d = 24 cm in-situ thermal insulation 035, d = 16 cm filigree slab	0.24	44356.– 427.–	97468.– 938.–	83.5	16.5	26	low	low
3. Mixed construction concrete walls + timber slabs and walls, thermal insulation 035	0.24		96627.– 928.–	83.4	16.6	22	middle	high
4. Expanded clay concrete BN 5 with expanded glass insulation solid concrete slabs	0.28		95928.– 924.–	83.3	16.7	16	middle	high
5. Aerated concrete walls λ = 0.12 and thermal insulation 035 with polystyrene			89880.– 866.–	82.2	17.8	18	low	low
6. Timber stud construction thermal insulation 035 with mineral wool and external render	0.18	46687.– 433.–	98683.– 916.–			6	high	high
7. Solid timber construction in panel construction system	0.18		111205.– 1036.–			6	high	high

All costs are total [incl. 19% V.A.T.] and incl. contractor surcharge. These costs are not legally binding.

tions all lead to unnecessarily expensive constructions. Innovation must be sought to remedy this fact.
Uneven brickwork, for example, can result in expensive installation work and considerable costs for plastering and rendering. Due to the additional plaster work, huge levels of moisture are introduced into the building and the necessary drying time causes additional extensions to the site planning.

The G.A.S. system house

The key to this construction system is with as few prefabricated, standardized elements as possible to achieve as many different housing combinations as possible. Individually manufactured components should be avoided or at least restricted to a minimum. Customer orientation can be including without incurring extra costs and it is possible for manufacturers to provide products of consistently high quality.
The underlying philosophy of system building also has to be understood by estate agents and communicated by them to prospective clients.
A further condition for economic forms of housing is the optimisation of building sites, which, in turn, requires local authorities to enter into dialogue with clients and planners. Although the system planning offers great variation and numbers of house types, fundamental decisions are often impossible to incorporate. For example, a decision regarding the window formats may not be compatible with the house dimensions and module sizes, or their placement on the allocated site. Therefore, the site and its infrastructure must also be optimised, when the aims are greater than simply constructing the most economical house. Experience with the relevant planning authorities in Germany has been positive to date. Particularly when the optimisation of sites has led to cost advantages which were directed back to the clients from the general contractors.
It is fundamental when building system houses to integrate all those involved in the planning and construction into a team. The architect should play the role of expert co-ordinator and should not hand over the constructional planning to other consultants. Through a combination of CAD and CAM, architects now have a means of involving themselves in the factory production of the elements they have designed. The design and planning of optimised elements represents and essential task in the constructional planning. Regrettably, this task was often handed over to others in the past.
The acceptance of system building, by clients and builders alike, is surprisingly large in Germany. The demands of the clients for a high quality product play a decisive role in this; successful quality control is more easily carried out in system building construction. No longer is the variety of lifestyle alternatives reserved for more financially stable clients. In view of the cost pressures to which housing construction is subject, it will only be possible with system building to implement the appropriate dwelling concepts for the different forms of households that exist in modern society. Experiments in must succeed now, in this decade, in order to give system building a meaningful role in the future (third attempt after the Bauhaus and other experiments in the 1950s and 1960s).

1 Placement of a prefabricated wall element
2 System house in Konz

Terrace Housing in Milton Keynes

Architects: Rogers Stirk Harbour + Partners, London

High quality living space at low cost
Passive house standard with low running costs

The creation of high quality living space at low cost was the task set by an open for passive energy housing. A further requirement was that the purchase prices of 30% of the constructed dwellings should not exceed £60,000. The solution conceived by the architects equally combines urban integration, passive house standards and a high standard of living. Located on the fringes of Milton Keynes, 60 km north of London, the site stretches over an area of 3 hectares and encompasses 145 dwellings, 56 of which remain below the £60,000 threshold. Additionally, 43 houses could be bought on the shared ownership programme; a section of the real estate being bought outright while the rest is rented and simultaneously paid off over a period of time.

Various housing types with maximum flexibility

This densely built development with interactive visual links is reminiscent of traditional village structures with squares and lanes, in spite of its contemporary appearance. Three circle roads access the development from the south and parking is spread between the houses in small-scale carports. All dwellings are threshold-free and based on the same elemental construction system. They are divided into two building halves which are off-set from each another. The slender prefabricated function and communication module accommodates entrance, corridor, stair and bathroom, and can be fully assembled within 31 days. The second module is flexible, in that users can adapt the layout according to their own personal requirements, and can be extended by additional modules. Windows and roofs can also be individually selected, producing nine possible combinations of two and three storey house types.

Optimised energy consumption

The flexibility of the concept allows various alternatives of building materials to be selected dependent upon the locally accessible raw materials. In order to achieve the required passive energy standard, it was decided to construct 25 cm deep external walls and to employ air-tight connections, thereby reducing transmitted heat loss levels to a minimum. Thus, it was possible to reduce the energy consumption of the buildings to 27% less than a conventional construction of comparable size. Heat is delivered via solar collectors which are located on the so-called "Eco-Hat", a component that can also be ordered. Exhaust air is collected and centrally transported to heat exchangers, where the residual heat is withdrawn and re-used to heat the in-coming air.

Strategies for cost efficiency

The decision to construct the houses in prefabricated, timber framed elements is essentially based upon the prerequisite that the project provide cost efficient living space. On one hand the construction time is reduced, on the other hand consistent quality standards can be secured. Wall and ceiling elements for the living and bedrooms were predominantly built on site, while the function modules were delivered ready for assembly. The facades are clad with a variety of different coloured panels in a range of sizes.

Project details:		
Usage:	Terrace houses 5 types from 57.5–102 m²	site plan scale 1:3500 floor plans · section house type C
Construction:	timber framework	scale 1:200
Internal ceiling height:	2.35–3.4 m	Isometric view
Total construction cost:	19,225,214 Euros	of modules
Built site area:	3,300 m²	
Period of construction:	26 months	
Date of construction:	2007	1 kitchen
		2 living
House type C:		3 WC
Total Cost:	88,731 Euros	4 bedroom
Cost per m² built area:	1,167 Euros	5 children's room
Total internal volume:	199 m³	6 bathroom
Total built area:	76 m²	7 study

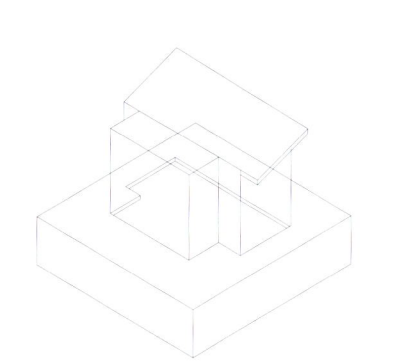

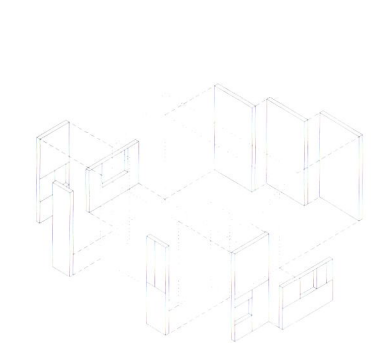

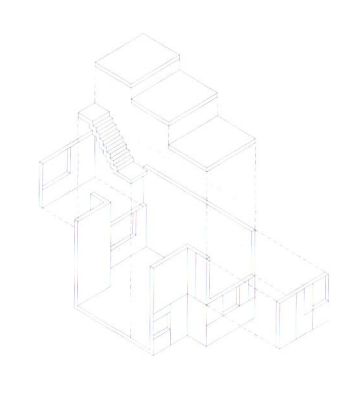

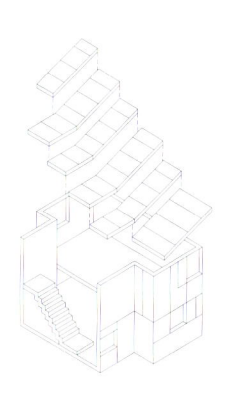

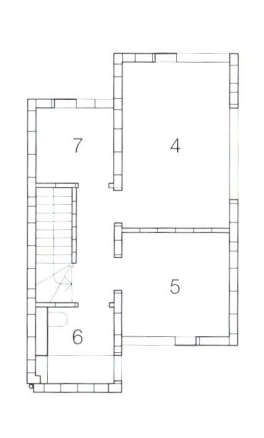

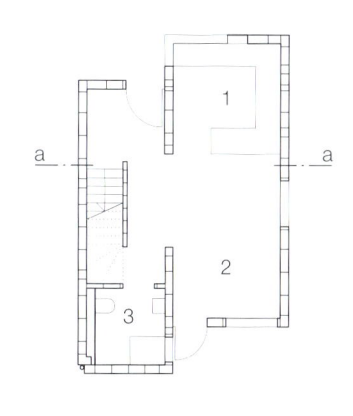

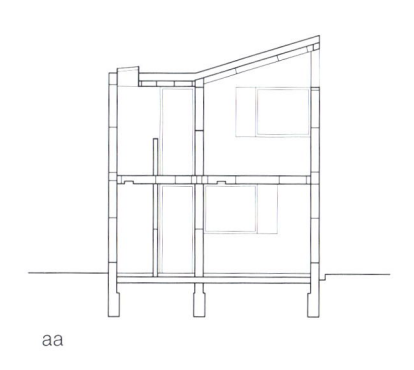

aa

71

Apartment House in Dortmund

Architects: Archifactory.de, Bochum

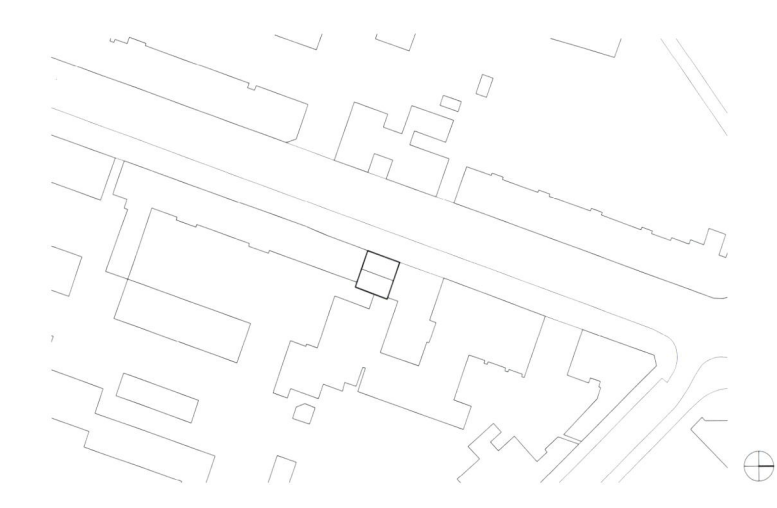

**Renovation and vertical extension
Increased density to increase efficiency
Stairwell functions as exhibition space**

The process of rehabilitating the Ruhr industrial area and turning it into a modern urban environment is clearly evident in the district of Hörde in Dortmund. Plans exist to transform derelict areas once occupied by blast-furnace works into a location for innovative activities, with a mixture of research and development, service and leisure facilities, as well as housing. A few streets away a similar concept has already been implemented in the conversion of a small apartment house. While the function has not been altered, the appearance has drastically changed. The outcome is a "new" clear-cut cubic structure with a traditional double-pitched roof which integrates well into the local urban fabric.

External appearance
On the street frontage, the building asserts itself with a clear presence; both the front aspect and the gable end (overlooking an access courtyard) have an urban significance. Partially set back from the broad pavement, the entrance is covered by a finely articulated glazed canopy. The windows have been set flush with the outer facade. The smooth rendered outer walls merge seamlessly with the roof, which is clad in zinc-sheeting and the gutters are integrated into the planes of the roof.

Internal organisation
The staircase is illuminated from above through a skylight, which is concealed in the rear slope of the roof, and via a light well. This dispensed with the need for windows in the corresponding facade areas. Replaceable aluminium panels with vibrant screen-printing steep the stairwell in coloured light, creating a welcome contrast to the monochromic appearance of the rest of the house.

The stairwell takes on the role of exhibition space for the owner. The original sanitary facilities and access core were removed during renovation. By putting two alternative stair designs simultaneously out to tender it was possible for the client to choose the most economical solution. Instead of a steel balustrade a solid spandrel winds sculpturally up the new concrete staircase.

Cost efficiency
Because the original roof was not to be retained, the opportunity to extend vertically presented itself, and while this decision was the result of both economical and urban considerations, it also enhanced the cost efficiency of the entire project. The floor plans of the two existing levels, accommodating one dwelling each, remained largely untouched and only the fittings were modernised. The new dwelling is planned around an open living and dining area.

Precise detailing and specification enabled the desired zinc-sheet roof to cost only minimally more than a traditional pan-tiled roof. In order to achieve a smooth, flat facade projections in the existing external walls were equalised by the application of two layers of a composite thermal insulation system. Although increased costs were involved in the application of the insulation, it was a more economical solution than the removal of projecting columns and pilasters, as well as providing significant reductions in future energy consumption.

site plan
scale 1:2000
sections · floor plans
scale 1:400

1 hall
2 bathroom
3 kitchen
4 room
5 living/dining

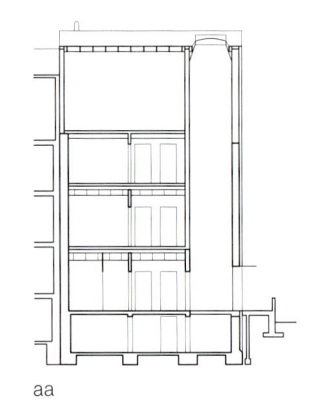

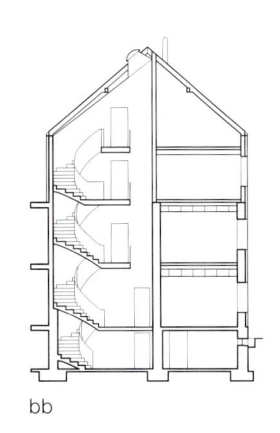

aa bb

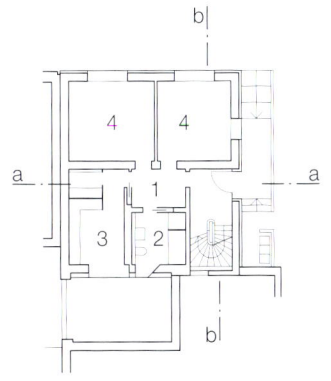

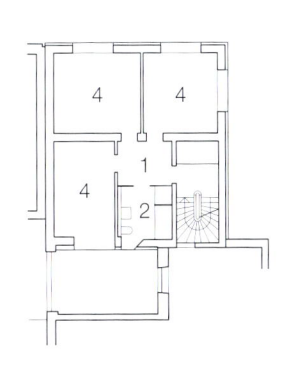

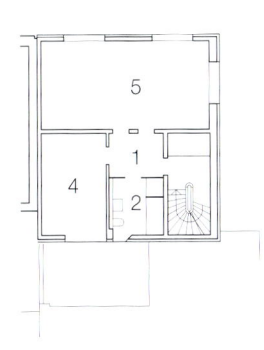

Project details:
Usage: apartment house
Units: 3 × 65 m² residential area
Net construction cost: 200,000 Euros
Net demolition cost: 10,000 Euros
Net renovation cost: 150,000 Euros
Net extension cost: 40,000 Euros
Total construction cost: 404,000 Euros
Cost per m²: 360 Euros
Total internal volume: 1,440 m³
Total floor area: 556 m²
Site area: 1,855 m²
Built site area: 110 m²
Date of construction: 2004

detail section
scale 1:20

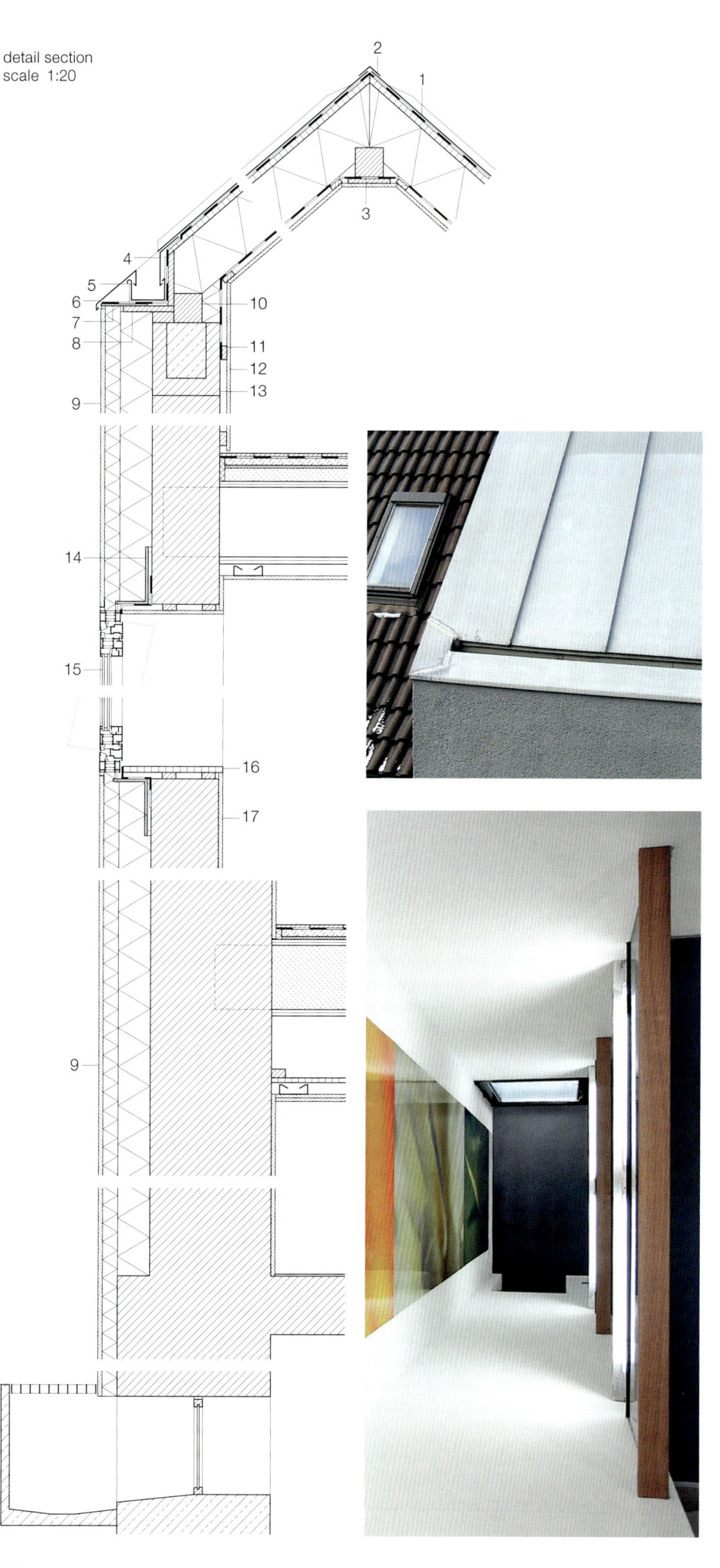

1 roof construction:
 0.7 mm pre-weathered sheet-zinc roofing with double-lock welted joints
 0.8 mm separating layer, 22 mm OSB
 180 mm mineral-fibre thermal insulation
 80 × 180 mm rafters, vapour barrier,
 24 × 48 mm battens, 12.5 mm plasterboard
2 sheet-zinc ridge covering
3 100 × 100 mm timber ridge purlin
4 0.7 mm sheet-zinc gutter flashing
5 sheet-zinc box gutter
6 sheet-zinc eaves flashing
7 1.5 mm sheet-aluminium closing strip
8 220 × 30 mm timber plate
9 wall construction:
 20 mm coloured scraped rendering
 160–180 mm two-layer composite thermal insulation system, 240–400 mm aerated-concrete blockwork, new
 240–430 mm brickwork, existing
10 100 × 100 mm timber eaves plate
11 24 × 48 mm timber battens
12 12.5 mm plasterboard
13 240 × 250 mm sand-lime U-section peripheral tie beam
14 150 × 200 × 10 mm aluminium angle
15 side-hung aluminium casement, stove-enamelled with double glazing:
 4 mm float + 16 mm cavity + 4 mm float
16 20 mm medium-density fibreboard window sill, painted white
17 15–20 mm gypsum plaster
18 1000 × 2000 mm domed skylight,
 2× 3 mm clear Perspex
19 fibre-reinforced polyester-resin base ring with thermal insulation
20 tapered timber bearers
21 vapour barrier
22 100 × 60 mm timber plate
23 sheet-metal verge covering
24 140 × 550 × 10 mm galvanized steel angle
25 laminated safety glass: 2 × 8 mm partially toughened glass + 0.76 mm PVB membrane
26 140 × 10 mm galvanized steel plate
27 5 mm elastomer seal on both sides
28 60 × 60 mm timber bearer
29 3 mm stove-enamelled aluminium sheet
30 stove-enamelled alum. entrance door with double glazing: 2× 4 mm float + 16 mm cavity
31 3 mm linoleum
32 180 mm reinforced concrete slab
33 60 mm concrete topping
34 liquid bituminous seal
35 200 × 100 × 10 mm galvanized steel angle
36 150 × 65 × 5 mm stainless-steel angle
37 250 mm reinforced concrete slab
38 40 mm thermal insulation
39 stove-enamelled letter box
40 aluminium frame: 2, 25 × 50 × 4 mm sections, thermally separated

75

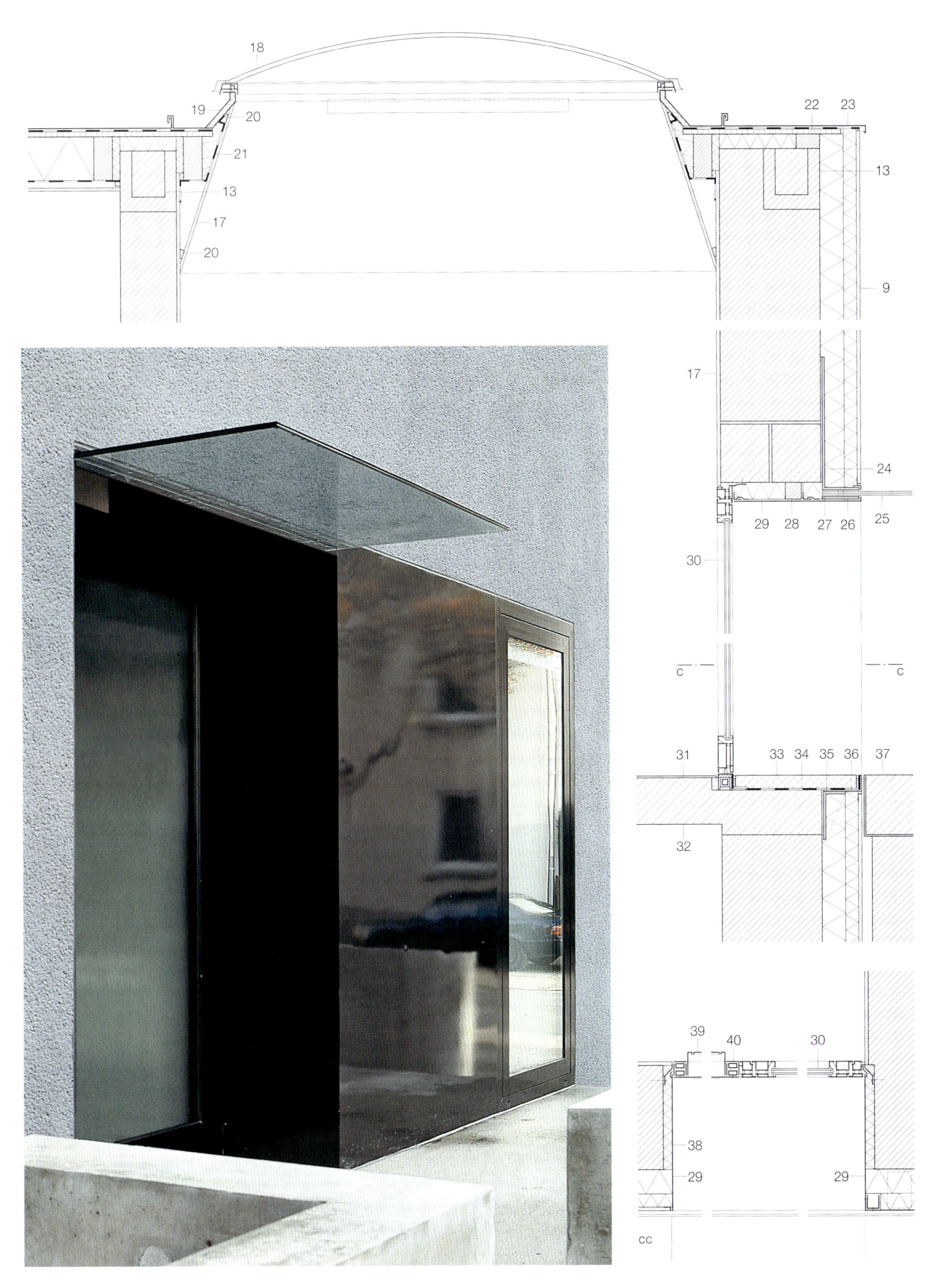

Housing in London

Architects: Ash Sakula Architects, London

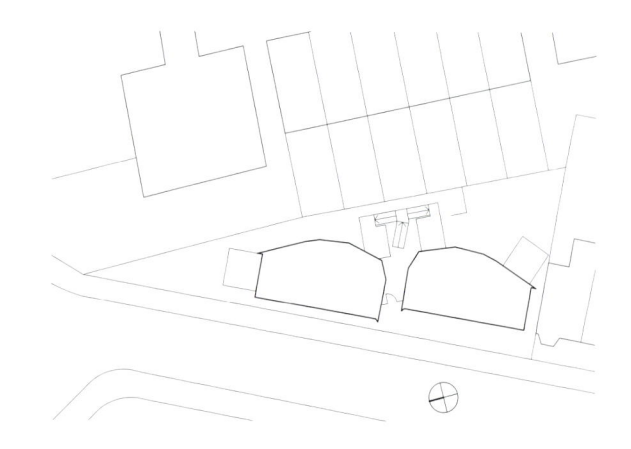

Aluminium membrane skin

This rather experimental project, accommodating four dwelling units of 69 m² each, is located amongst an assortment of otherwise characterless housing blocks. Silvertown, in London's east near the Thames, is presently one of the most popular new residential areas with numerous housing projects being built. Just how attractive and innovative low-budget housing can be is demonstrated by this eye-catching project by the young architectural firm of Ash Sakula. The scheme is the result of an architectural competition sponsored by the Peabody Trust, which regularly organises competitions for experimental architecture.

Visible facade construction

The experimental aspect of the project manifests itself above all in the facade, which consists of an aluminium foil protected by a yellow, transparent fibreglass rain-screen. The corrugations of the panels run horizontally on one face and vertically on the other, with twisted, recycled wire elements – created by the artist Vinita Hassard – mounted beneath the corrugated panels.

Innovative plans

The layout is also atypical with the spacious entrance hall dominating the 69 m² dwelling. Fitted with generous cupboards and storage space, it can be used as a workroom, play space or utilities area. The comparatively small living room acts as a refuge and place of relaxation, TV room or guest room. The kitchen is generously dimensioned and offers a central space for the entire family.

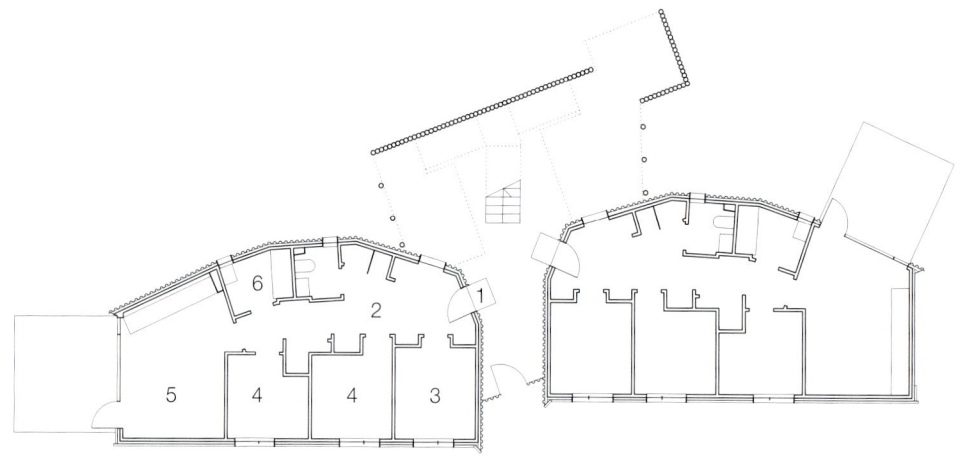

site plan
scale 1:750
floor plan
scale 1:250

1 entrance
2 hall
3 TV/guest room
4 bedroom
5 kitchen/living
6 bathroom

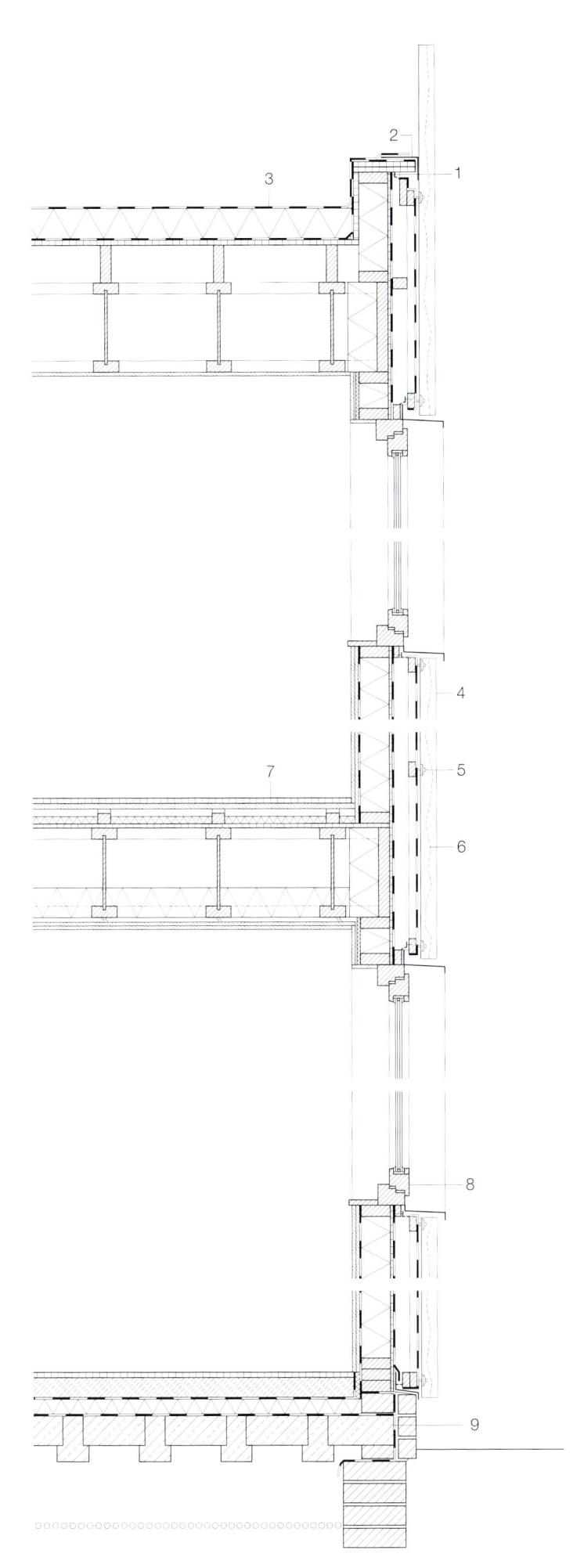

vertical section
scale 1:20

Project details:
Usage:	Housing
Construction:	Timber framework
Internal ceiling height:	2.38 m
Construction cost:	865,830 Euros
Cost per m² area:	2202 Euros
Total internal volume:	918 m³
Total built area:	338 m²
Area per dwelling:	69 m²
Usable area:	268 m²
Built site area:	303 m²
Total site area:	472 m²
Date of construction:	2004
Period of construction:	12 months

1 insect screen
2 sheet-aluminium, bent to shape
3 roof construction:
  two-layer roof seal
  100 mm thermal insulation
  vapour barrier
  18 mm plywood
  prefabricated timber-frame construction
4 wall construction:
  151 × 77 mm corrugated glass-fibre-reinforced, polymeric polyester sheeting
  aluminium-foil heat-reflecting membrane
  25 × 50 mm counterbattens
  33 × 50 mm battens
  permeable membrane, stapled
  13 mm plywood
  100 mm thermal insulation
  vapour barrier
  2× 12.5 mm plasterboard
5 bolt fixing in corrugation, with washer bent to shape and elastomer compression rings
6 twisted, recycled electric cables
7 floor construction:
  18 mm compressed chipboard
  19 mm sound-absorbing panels
  50 mm battens, 20 mm insulation
  15 mm oriented-strand board
  prefabricated timber-frame construction
8 softwood window frame
9 prefabricated reinforced concrete floor elements

Apartment Building in London

Architects: Niall McLaughlin Architects, London

High expectations on a low budget
Building skin evokes local history
Economic dwellings on time-purchase concept

This building in Silvertown, in East London, does not look at all like a typical low-cost development. Yet this striking block of apartments was, in fact, realised on a tight budget. The title of the design competition, "Fresh Ideas for Low-Cost Housing", was the inspiration for the young architectural practice of Niall McLaughlin and Associates. Low-cost housing in England usually takes the form of pre-fabricated timber construction, concealed behind a traditional cladding of timber or brickwork. In this case, the experimentation began with the facade: low-cost, iridescent, gift wrapping film was enlisted for a unique function. The shimmering colours are a reference to the history of the location: up to the 20th century, factories were manufacturing cheap luxury goods there, from dyes to confectionary.

Large internal openings influence the spatial experience and frame the views of the surrounding urban landscape. The layouts are functional and unspectacular and the walls are simple, clear white – thereby giving the residents the freedom to make their own mark on their dwellings. Each living unit has two bedrooms and a bathroom. The kitchen, dining and living spaces merge together and are accommodated on the southern side of the building. There is a small loggia or terrace included in each flat and the ground floor apartments benefit from back gardens.

In allocating the economical yet attractive three-room apartments, priority was given to local residents registered in the Newham area. Tenants from community or other non-profit housing associations were also deemed eligible for the dwellings. One pre-requisite did however exist: an annual income of £28,758 for singles or £32,644 for couples. This was necessary in order to ensure the success of the particular programme of shared ownership which was implemented; a relatively new concept in England that foresees the purchase of property over time. In order to obtain a loan from a bank a minimum income is necessary. Part of the property is sold; in this case 30–75 per cent, while the rest is initially rented out and purchased by the tenants over a period of time. The purchase price for the 75 m² flats was £210,000, which corresponded with a capital investment of at least £63,000. The logical, well laid out dwellings met with great response; all dwellings had been sold by the beginning of 2005.

Iridescent film

The architects developed a concept for the south facade of the apartment building in collaboration with the artist Martin Richman. Polycarbonate louvers are centrally set within 200 mm deep aluminium trays behind sheets of cast glass. The louvers and the rear surfaces of the trays are covered with strips of gift wrapping film. Light falling on the facade is reflected back from the various layers, producing a shifting pattern. The external layer of cast glass further breaks up the light as it is reflected back from the facade panel. As the angle of incidence of the light changes, so does the play of colours in the facade, due to the various reflective qualities of the film. The ingenious interaction of commercially available wrapping film, light and the observer himself continually leads to new visual effects.

Cost efficiency

It was possible to secure considerable cost savings through the intensive coordination between the various project participants, and the application of pre-fabricated construction elements. The building was erected on in-situ concrete piles and strip footings and the ground floor slab consists of pre-fabricated reinforced concrete elements. The walls and floors, based on a panelled timber-frame construction, and the staircase were all prefabricated. Further cost savings were achieved by restricting the project to a limited number of window sizes. The main technical problems were presented by the integration of the large, heavy glass facade elements into the framed construction of the southern facade, as well as the corner windows and the large openings for the staircases. To overcome these problems, supplementary steel beams were inserted into the timber framed construction.

Project details:
Usage:	single storey dwellings 12 dwellings of 79–82 m²
Construction:	timber framework
Internal ceiling height:	2.41–2.98 m
Total construction cost:	2.2 million Euros
Cost per m² usable area:	2,200 Euros
Total internal volume:	2,870 m³
Total built area:	1,071 m²
Residential area:	966 m²
Site area:	1,013 m²
Date of construction:	2004
Period of construction:	15 months

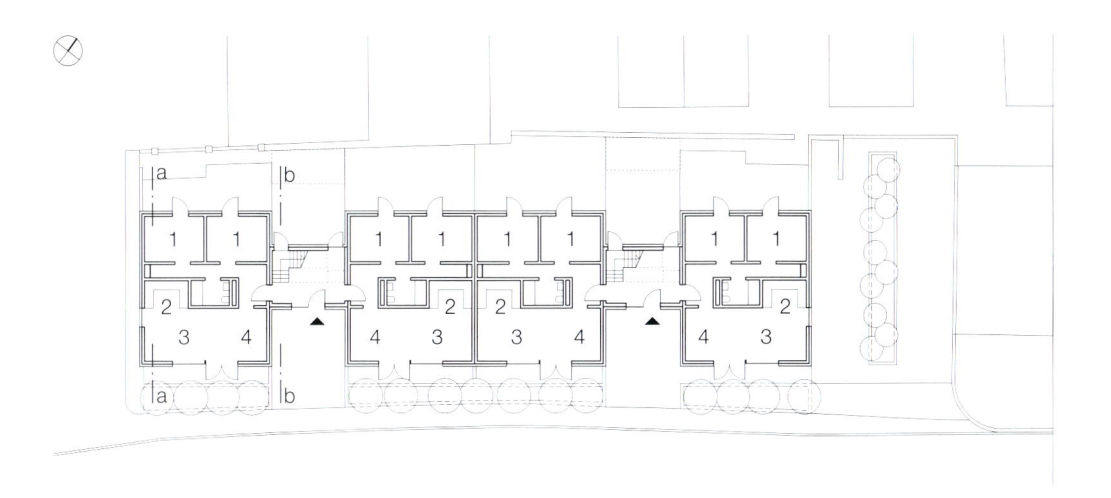

floor plans
sections
scale 1:500

1 bedroom
2 kitchen
3 dining
4 living
5 loggia

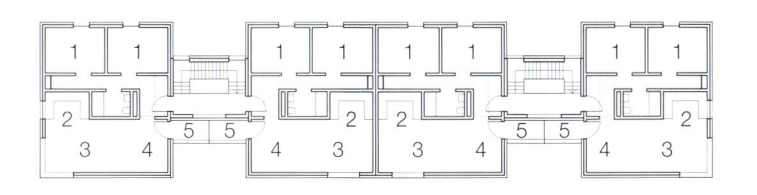

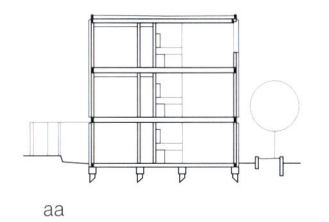

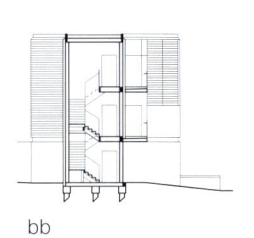

vertical sections
horizontal section
scale 1:10

1
2
3
4
5

80

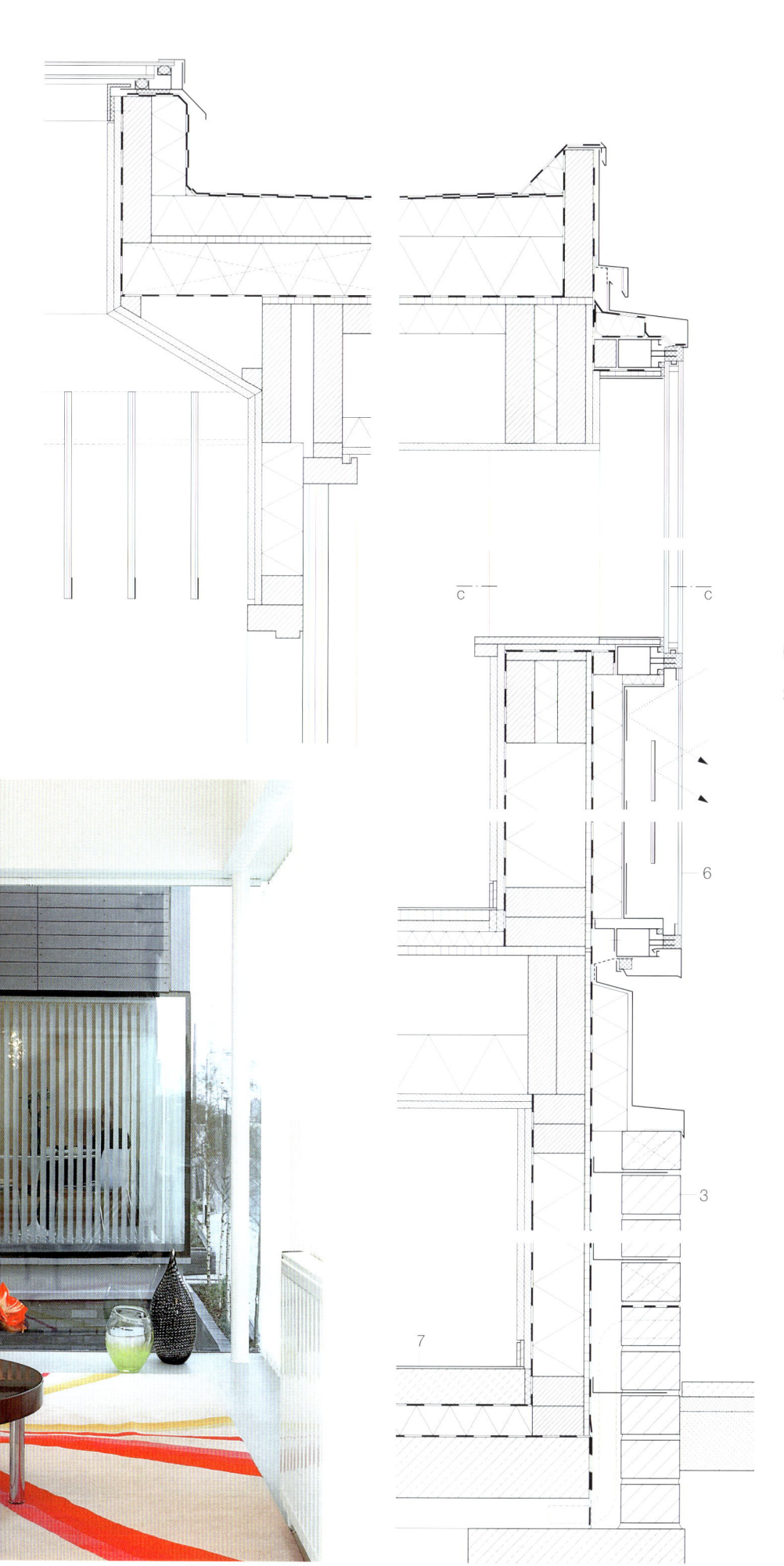

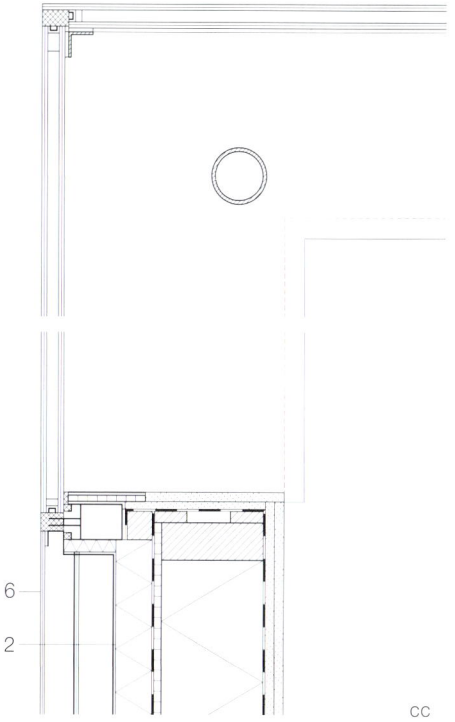

1 24 mm larch boarding
2 wall construction:
   fire-resisting sheeting,
   50 mm ventilation cavity
   windproof layer
   10 mm oriented-strand board
   140 mm thermal insulation between
   50 × 140 mm timber studding
   vapour barrier
   2× 12.5 mm plasterboard
3 clinker face-brickwork
   52 mm ventilation cavity
   windproof layer
   10 mm oriented-strand board
   90 mm thermal insulation
   vapour barrier
   2× 12.5 mm plasterboard
4 upper floor construction:
   carpet on
   18 mm plywood
   19 mm fire-resisting sheeting
   25 mm rigid-foam insulation
   15 mm plywood
   100 mm thermal insulation between
   50 × 235 mm joists
   2× 12.5 mm plasterboard
5 polycarbonate strips,
   lower areas with adhesive-fixed
   iridescent film
6 2150 × 1100 × 100 mm facade panels:
   aluminium tray with side slits for fixing,
   6 mm polycarbonate louvers,
   louvers and inside of tray lined with
   adhesive-fixed iridescent film
   6 mm cast-glass front with
   adhesive-fixed UV-filter film, internally
7 3 mm synthetic-resin layer
   60 mm cement screed on polyethylene
   membrane, 50 mm rigid-foam insulation,
   sealing mambrane
   155 mm reinforced concrete floor elements

vertical section   scale 1:10

1   2150 × 1100 × 100 mm facade panels:
    aluminium tray with side slits for fixing,
    6 mm polycarbonate louvers,
    louvers and inside of tray lined with
    adhesive-fixed iridescent film
    6 mm cast-glass front with
    adhesive-fixed UV-filter film, internally
2   fire-resisting sheeting, 50 mm ventilation cavity
    windproof layer, 10 mm oriented-strand board
    140 mm thermal insulation between
    50 × 140 mm timber studding
    vapour barrier, 2× 12.5 mm plasterboard
3   carpet on 18 mm plywood
    19 mm fire-resisting sheeting
    25 mm rigid-foam insulation, 15 mm plywood
    100 mm thermal insulation between
    50 × 235 mm joists, 2× 12.5 mm plasterboard

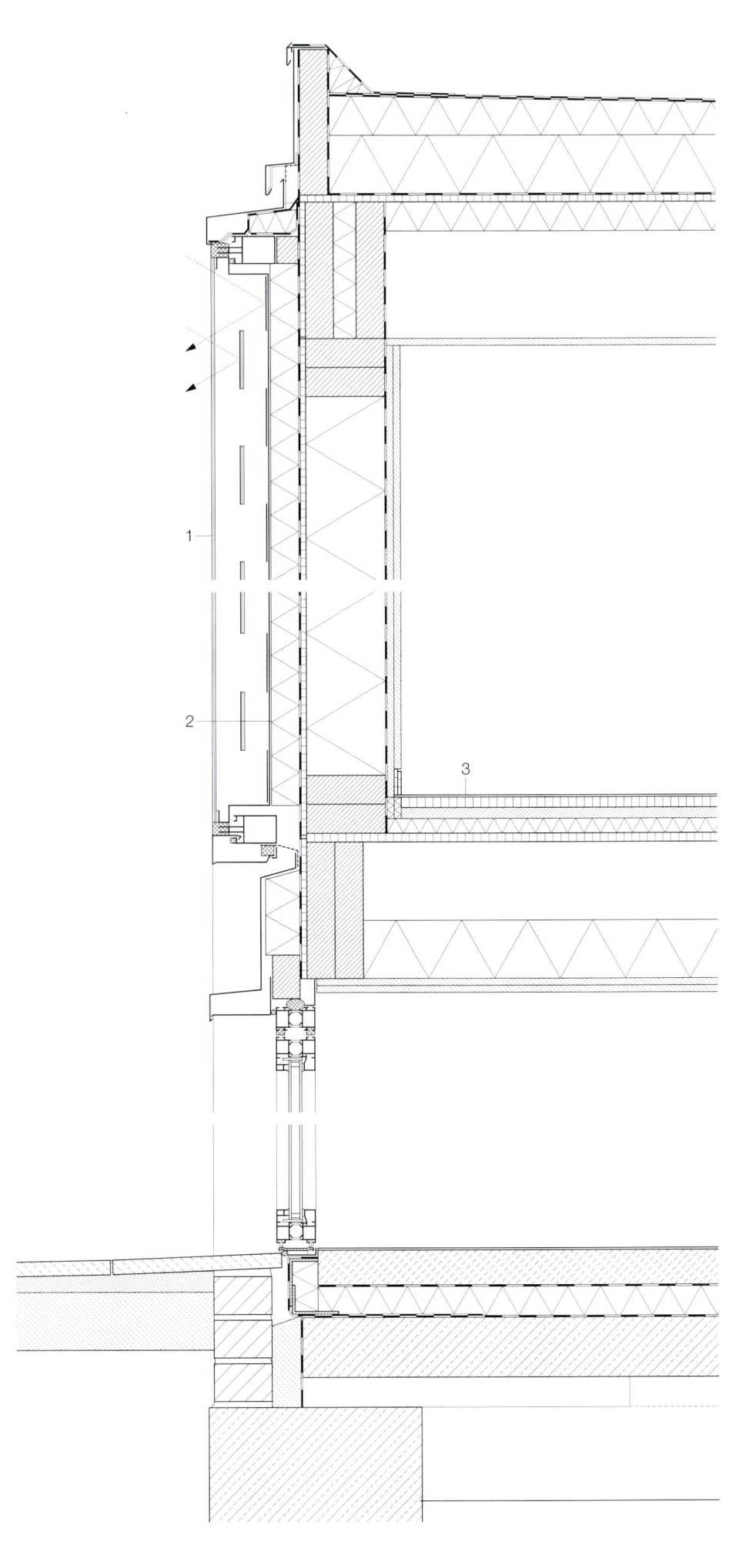

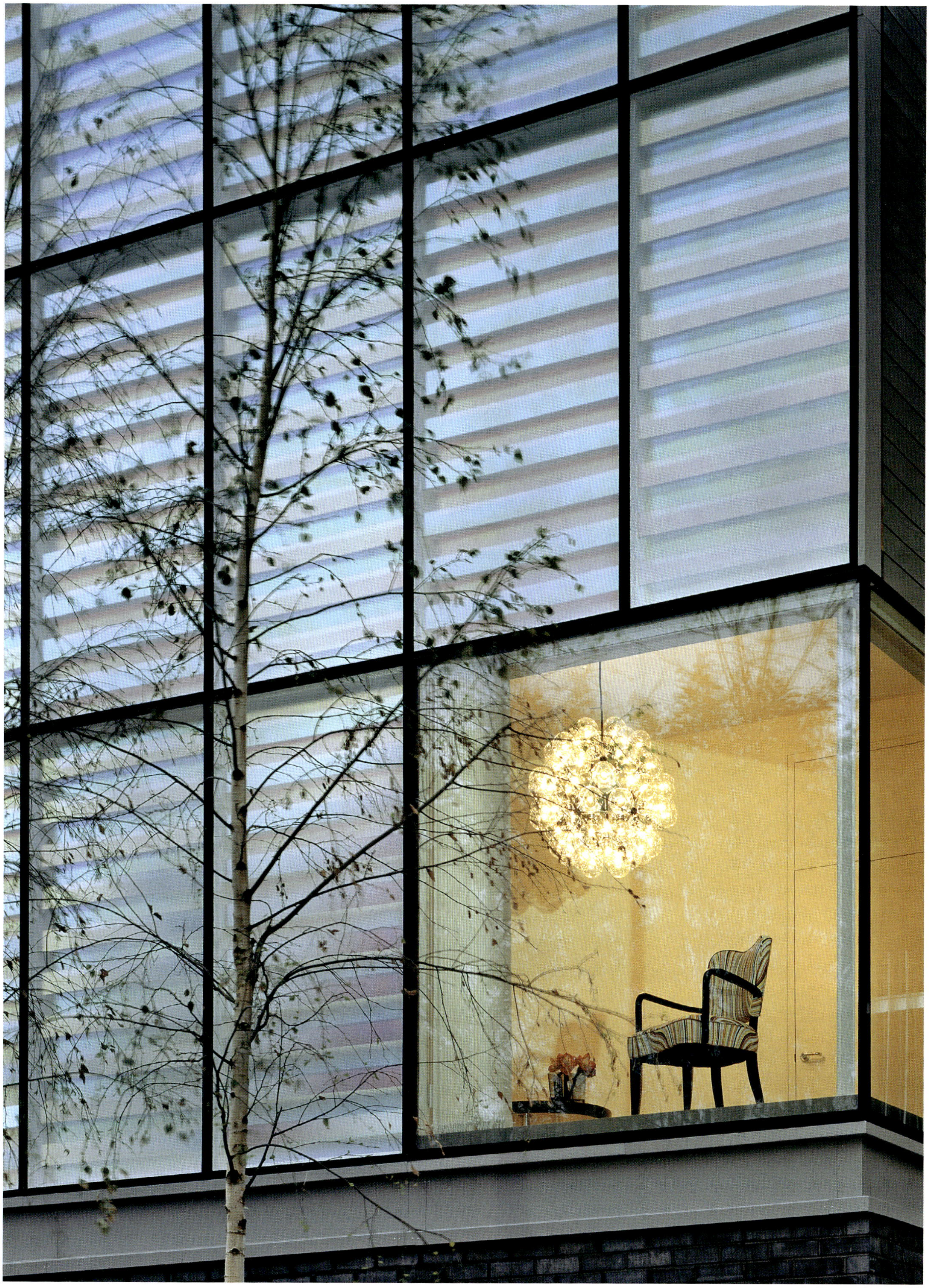

Multi-Storey Housing in Munich

Architects: Hierl Architects, Munich

25 different dwelling types
Finalisation of detailing prior to construction
Naturally ventilated underground car parking
High quality residential space for 1,149 Euros/m²

Urban planning policy in Munich today decrees an increase in the density of building development in the city centre. This housing development on Marlene-Dietrich-Strasse in the Arnulf Park district is part of this programme. The reclamation of former railway yards between Munich's central station and the suburb of Pasing resulted in a generously sized redevelopment quarter to be made available for office and residential space. Alongside what are described as "luxury" flats, 106 publicly funded social dwellings have been constructed.

Attractive blend of dwellings

25 different dwelling types, ranging from 37 m² apartments to 130 m² five-room dwellings, cater to a wide variety of residential demands and provide simple yet high quality living space. All detailing alternatives were finalised prior to the start of construction, from the selection of floor finished to the detailing of the bathrooms. This enabled cost variations to be kept to a minimum and the comparatively modest construction budget to be adhered to.

Economical facade treatment

The strict rhythm of the street frontages of this triple-sided peripheral construction is treated with an anthracite-grey, composite thermal-insulation system that appears to shimmer depending on the light. Ceiling high flush-set windows and inserted loggias articulate the almost 100-metre-long southern facade. In contrast, the facades to the more private spaces, the courtyard and loggias, are finished in a white render softened with a nuance of pale-green.
Changes in height and the play of colour of the anthracite and red window frames determine the appearance of the courtyard facades. Cost savings could be achieved through the use of consistent windows sizes in the vertical; for the street frontage, and in the horizontal; for the courtyard.

Cost savings in the basement level

Savings were achieved by designing the basement garage to rise somewhat out of the ground and by dispensing with a second basement level. Excavation could be carried out without the need for waterproofing against the ingress of groundwater, and tanking of the basement was also unnecessary. Natural north-south ventilation of the underground car parking facilities is provided via facade openings which further contributed to cost savings. Storage spaces are provided on the northern side of the roof storey, replacing those normally accommodated in a basement level.
The savings made in the technical budget benefited the actual dwellings. As the site progressed, the client decided to upgrade floor finishes in all living spaces, with the exception of the kitchens, from the originally specified linoleum to industrial quality parquet. As a result of various cost-saving strategies it was possible to create high quality housing with relatively low construction costs; 1,149 Euros per m² including parking.

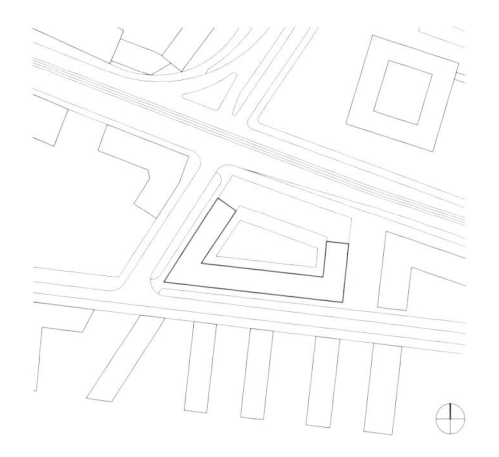

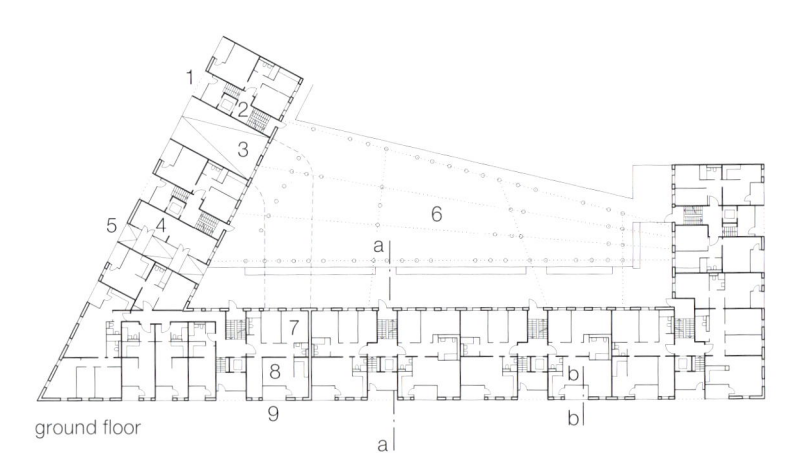

ground floor

floor plan types
scale 1:500

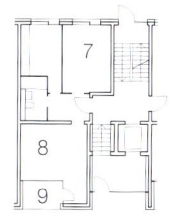

A West tract:
2-room dwelling
54 m²

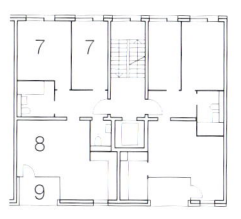

B East tract:
3-room dwelling
60–68 m²

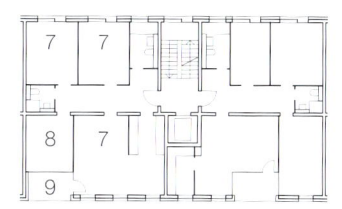

C South tract:
4-room dwelling
92–95 m²

Project details
Usage: Housing development
Total construction cost: 8.729 million Euros
Cost per m² residential area: 1,149 Euros (incl. parking)
Cost per m² basement area: 982 Euros
Total internal volume: 35,371 m³
Total floor area: 10,450 m²
Residential floor area: 7,595 m²
Site area: 4,939 m²
Date of completion: 2005

site plan
scale 1:4000
section · floor plans
scale 1:1000

1 entrance
2 staircase
3 garage ramp
4 store
5 walkway
6 internal courtyard
7 bedroom
8 living/dining room
9 loggia

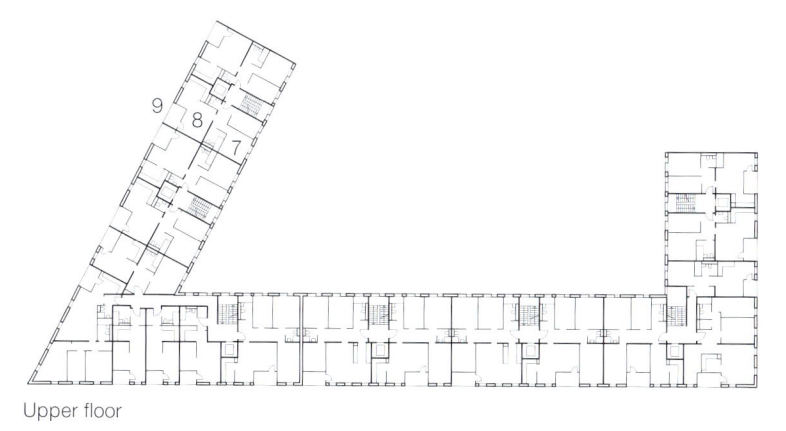

Upper floor

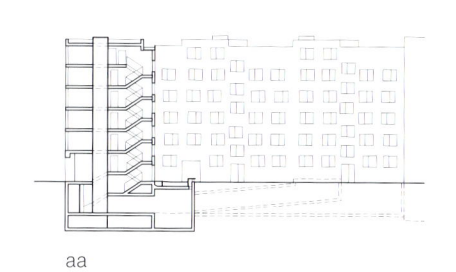

aa

bb

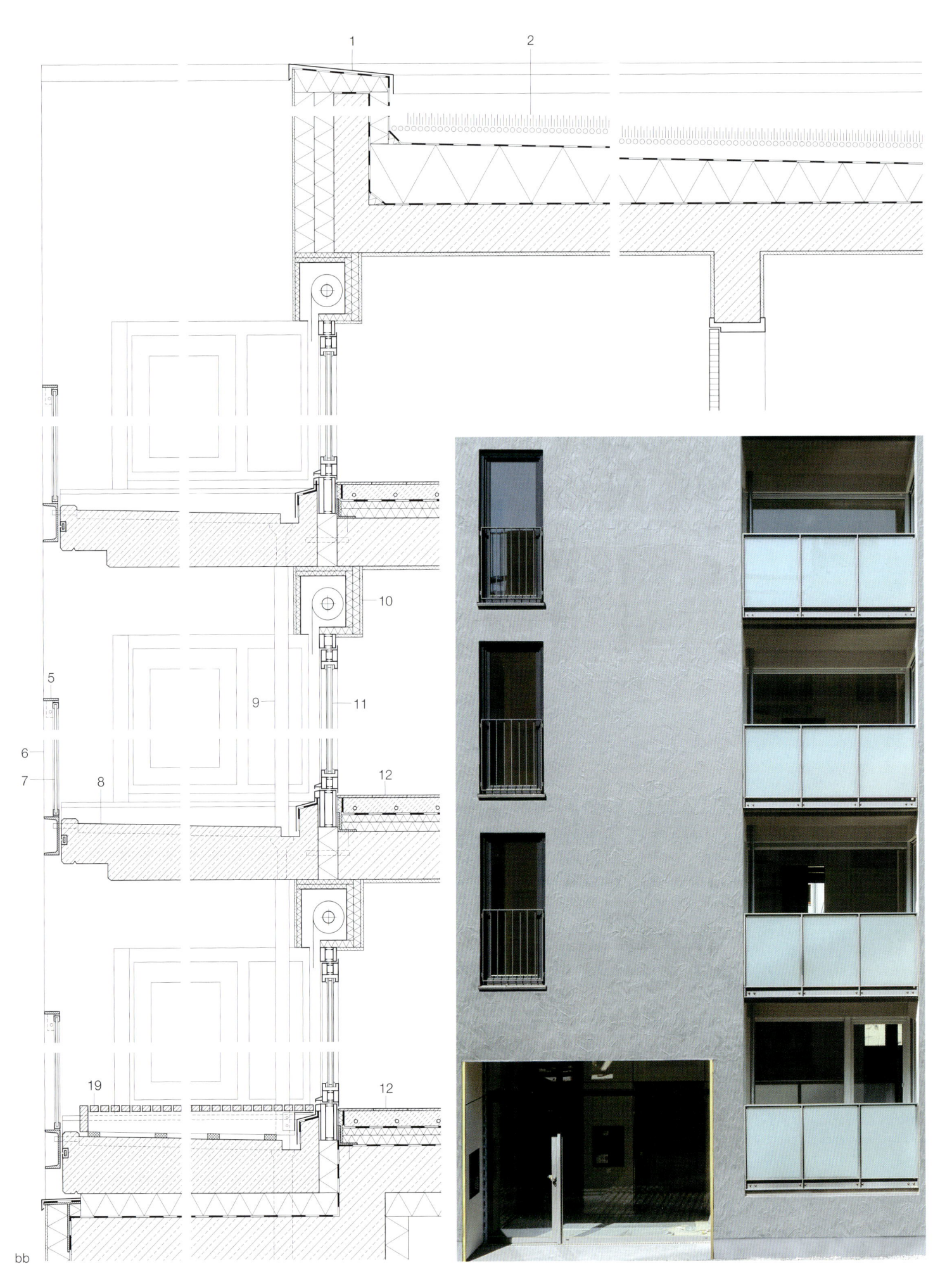

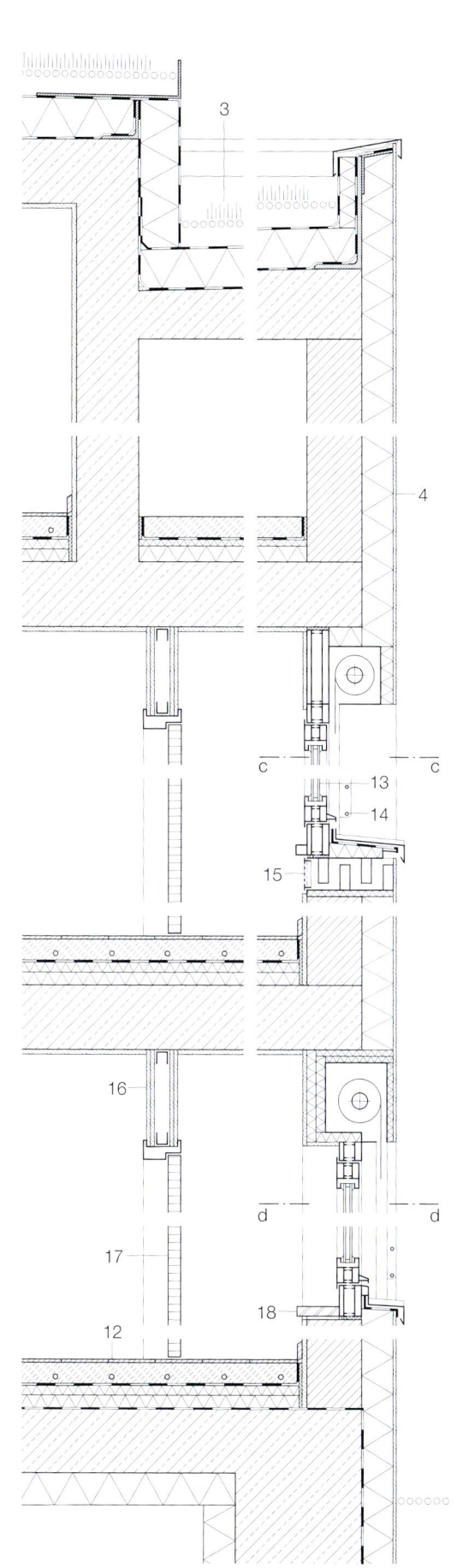

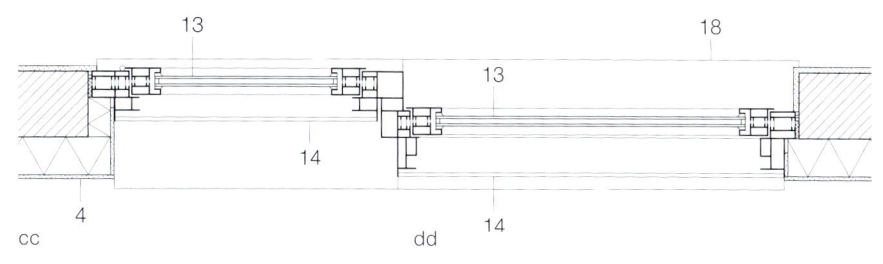

vertical section · horizontal section
loggia · courtyard
scale 1:20

1 parapet construction:
   2 mm powder-coated aluminium sheet
   3 mm double layer bituminous membrane
   60 mm insulation, 2 mm vapour barrier
   160 mm reinforced concrete
2 roof construction:
   80–100 mm planted layer
   3 mm double layer bituminous membrane
   250–100 mm polyurethane rigid-foam
   insulation, 2 mm vapour barrier,
   200 mm reinforced concrete slab
   8 mm render, painted white
3 80–100 mm planted layer
   1 mm separation layer
   120 mm extruded polystyrene insulation
   3 mm double layer bituminous membrane
   200–140 mm reinforced concrete slab
4 wall construction:
   silicone-resin finishing coat, rush-green
   to courtyard, anthracite-grey to street
   weather-resistant pure acrylate colour coat
   moisture-permeable moulded rendering
   6 mm adhesive layer reinforcement
   100 mm polyurethane rigid-foam insulation
   180 mm brickwork
5 60 × 15 mm RHS galvanised powder-
   coated flat steel handrail
6 60 × 15 mm RHS flat steel posts
7 6 mm opaque laminated safety glass
   balustrade
   15 × 15 mm steel channel frame
   120 × 400 × 8 mm steel channel beam
8 loggia construction:
   260–220 mm reinforced concrete slab
   thermally separated reinforcement
   connection
9 Ø 60 mm aluminium drainpipe
10 300 mm roller-shutter box, rendered white,
   with sound and thermal insulation
11 double glazing in plastic frame:
   4 mm toughened glass
   + 16 mm cavity + 6 mm toughened glass
12 floor construction:
   10 mm industrial quality parquet
   65 mm screed with underfloor heating
   separation layer
   30 mm impact sound insulation
   40 mm polyurethane rigid-foam insulation
13 double glazing in plastic frame
   8 mm laminated safety glass + 16 mm
   cavity + 16 mm toughened glass
   2 mm aluminium window sill, bent to fit
14 safety rail to 4th floor/roof storey:
   Ø 15 mm stainless-steel rods and plates
15 Ø 125 mm sound-insulated ventilator
16 lightweight partition:
   40 × 40 mm steel channels
   2× 5 mm plasterboard to both faces
17 40 mm timber veneer door in steel frame
18 30 mm slate window sill
19 25 × 35 mm larch grid,
   40 × 80 mm timber framing
   80 × 90 × 5 mm aluminium angle,
   35 × 15 mm neoprene bearer

Cost and Quality Awareness in Construction

Rudolf Hierl

People love getting things on the cheap. A new DVD player costs only half as much as having the old one repaired (though admittedly, the old one lasted twice as long as the new one will). At the same time, although the prices of many things are tumbling today, people are also willing to pay through the nose for exclusive luxury items – an expression of the growing polarization that is taking place in society. Cost awareness has different parameters: the price alone is one factor and ultimate quality is another. In architecture, discussions seem to revolve around the former at present, yet still coupled with a desire to ensure adequate quality. The present article seeks to explore the relationship between the two: cost awareness in construction – without building cheaply.

Historical view

In Germany, cost awareness in building dates back to the prosperous 1980s and postmodern architecture. It was at the end of this era that the first endeavours were made to build in a cost and space-conscious manner. Studying the developments that were realized in that period, one is surprised at their visual attractiveness. These were achieved with a wealth of materials, details, surface finishes and forms, not all of which were sustainable, however. The reunification of the two parts of Germany in 1989 resulted in a huge shift in the political landscape. Linked with this, was an interweaving of financial realms that determined many economic processes, known today as "globalisation".

All this obviously had an effect on the building sector, which over the centuries had been relatively sluggish and not very innovative. Cost pressures and quality constraints have led to marked changes in building production. The planning and construction phases as well as the whole question of regulations have major influence on construction costs, and only by taking them into consideration can cost awareness in construction be possible.

1. Parallel Planning

One knows from experience that when the planning process is complete, but before construction begins, confidence is high that an economical structure can be built. One is also aware that things that have not been fully thought out during planning often reveal shortcomings in the end. This is evident especially in the relationship between the carcass structure and mechanical services. Recognition of these faults during the construction phase can necessitate a reconsideration of the planning and far-reaching measures on site, such as alterations to sections of the work which have already been completed. Steps of this kind – cutting of finished concrete, relocation of service ducts or rooms, structural reinforcement – when the scheme is at an advanced stage, are likely to be disproportionately costly and lead to additional planning requirements. Rechecking spatial relationships, floor plans, technical services, and standards - particularly those concerning noise, fire and thermal protection – in addition to a general increase of the scope of the project and an extension of the cost factors are the unfortunate results. Any changes will also have to be coordinated with other planning teams and with the contractors. In all likelihood, this will occur under time pressure, which is a further potential source of errors. In other words, planning work that occurs parallel to the construction process does not lead to the savings in time or money that it may at first seem to promise. Rather it often leads to difficulties in cost control, and holds back the building process after a particular point in time, as well as causing the logistics of the site to become more expensive. Once construction has commenced the volume of information explodes dramatically; in this age of CAD, internet, virtual coordination and paper-free planning one would assume that this would decrease rather than increase, and become easier to control. Alas it is not uncommon for critical decisions to be overwhelmed by this sheer volume of information, without planners immediately realising the consequences. These "sleepers" appear, often months later, and wreak havoc. Of course, the situation is now drastically different from that at the time of project begin; the contract is awarded, and negotiation control is predominantly in the hands of the employed planners. Alterations and extension in the planning provide welcome opportunities for financial readjustment of fees and costs.

Prior planning
In cases where the planning has been completed before construction begins, one can assume that all aspects of the works will have been considered. The building companies are responsible for the efficient planning of workshop and assembly facilities and, in order to ensure their own financial success, have great interest in locating and resolving any confusion or inaccuracies in planning. A comprehensive planning programme that has been finalised before the commencement of construction is a reliable basis on which to proceed with the work. Any risks in calculations will then lie in the execution of the work, not in the planning.

Building as a process
The call for planning to be completed before embarking on construction is based on an old concept of building as a process, the quality of which is founded in a universal design concept and that comes to full artistic fruition in ad-hoc decisions on site. The concept is based on a successive sequence of trade contributions. But neither of these situations exists any more. The old building process has been replaced by a complex division of labour and simultaneous production – often in the form of prefabricated elements. That calls for greater conceptual and design abilities on the part of architects in order to handle all the spatial ramifications. At the same time, it also requires an elaborate technical infrastructure within which the works can be executed. This compaction of the planning and construction processes affects the

relationships and interaction of the building elements with each other.

Building as a componential process
The construction process today is based on a fine division of responsibilities in which components are prefabricated parallel, yet independently of each other, before being simultaneously assembled on site. During the erection of the carcass structure, for example, the building skin might be manufactured as a series of facade elements, the assembly of which must take into account permissible constructional tolerances. One is speaking here in terms of 3 cm per 10 m, which indicates a clearly perceptible coarsening of the structure as a whole. That makes it pointless, for example, to draw up final tile-layout plans for sanitary areas before the carcass is complete. If a pre-planned dimensional coordination does exist, one can adhere to it only by using prefabricated sanitary cells (or alternatively by accepting elaborate on-site adjustments).

Simultaneous building
A simultaneous building process requires larger tolerances than chronological construction; it also increases the problems in transitional areas. Greater tolerances lead to a larger number of joints, which can mar the aesthetic effect. The aesthetics of prefabrication may, in fact, represent the component characteristic of construction, but does not often meet with popular acceptance.
If the architect wishes to implement an integral, cohesive design for a building, one of the greatest challenges lies in overcoming the problem of tolerances. If this is not resolved in the planning, or not fully understood by the supervising site engineers, cost complications will be unavoidable. All the advantages of prefabrication, of parallel production processes and simultaneous assembly will be eroded by adaptation measures, with large amounts of corrective day-work. Although jointing systems are commonly used to remedy such situations, they are the poorest response. Two diametrically opposed solutions are conceivable: joints can be integrated in the structural articulation of a building (e.g. behind a pillar or parapet wall or at a junction between floor finishes); or one can make a feature of joints as a design element.

High tech, low tech
Globalisation has lent building sites an international aspect and has greatly speeded up the whole construction process. Construction firms are modern nomads, moving from one site to another with their metal site offices, providing specialised services. This internationalisation reflects an ancient problem: the dependence of building activities on the relevant social context. While the elite constructions of the Greeks could develop high-tech know how in a compact, relatively controlled framework and refine it to the point of almost breathtaking perfection, the globally expansive Roman builders had to rely on more simple, robust and – importantly – widely comprehensible building techniques. The standard was, of necessity, low tech.
In the past, as today, globalisation has stimulated exclusivity in neither technology nor details. Its most positive contribution has been to reduce the price of mass-produced technical artefacts. This is a situation which leads to irritation in a country where construction techniques and building materials are still dominated by handwork skills. Such products from the building market are juxtaposed with craft construction: industrially manufactured windows, for example, with hand-built window sills, and both integrated into a holistic planning process. This sometimes leads to insoluble conflicts, so that architects must either address the quality of the products from the building market by designing them themselves; or they must use higher-quality products.

Interfaces
A further factor lies in the demarcation of the different areas of planning. When building production takes place simultaneously in parallel processes, cost awareness in the planning means taking more constructional interfaces into account. Standard procedures are not always the best support in this respect. There is often a division of competence between the architect and the specialist engineer. Greened flat roofs, for example, are designed by the architect up to the level of the sealing layer, from the root layer upward by the landscape architect, while lightning rods and other technical aspects are the domains of various consultants – all detailed into a single section. Each and every forgotten interface is a further money pit, generating expensive alterations and causing the savings potential to be drastically diminished.

Logistics
One final aspect of cost awareness in planning is the role of site logistics. Less is not just more; it is everything. Instead of optimising individual construction units, it is more economical to use a few robust elements: a constant floor thickness, for example, even if it may be over-dimensioned in some areas. Similarly, the specification of uniform brick and concrete qualities allows last-minute deliveries to sites where there is little storage space, even if only minor savings are actually made. Decisions made in favour of tolerant floor finishes and unified slab constructions, compared with individual variations in screed and impact-sound insulations dimensions or quality do, however, provide flexibility for subsequent alterations on site. Apart from general planning strategies, completed planning policy and in-built flexibility, there are always certain elements of financial significance. Notably, decisions affecting these matters have to be taken at an early stage of the planning – usually in the preliminary design phase – and it is difficult to change them at a later date. This applies in particular to matters affecting the carcass structure and the facade, but also to the integration of elements. One would think that it were self-evident that the effort invested in cost-control would be remunerated. The time invested in research

and planning is considerably greater than appears in any written specifications, and the effort involved in planning economically much greater than when cost bears little or no relevance.

2. Building structure

Although almost any constructional contortion is technically feasible today, that in no way diminishes the cost relevance of logical building structures. Clear-cut design principles are of enormous economical relevance, e.g. solid cross-wall or spine-wall construction, strict axial grids in skeleton-frame structures, walls situated above each other that do not have to be supported by trimming construction and constructions with few or no projections. That does not mean, of course, that in the transition from an upper floor to a basement garage, the structural system may not be changed. This, in fact, is sensible because a higher wall construction is deemed necessary for the landscaping and it is the reinforcement that is expensive. What drives up costs, however, is the sum of the individual or exceptional measures; here an extra beam, there a change in construction, and so on. On the other hand it is possible, particularly in housing construction, to ensure a maximum of free wall space for fit-out during the carcass construction stage by considering floor surfaces, underfloor heating and door fittings.

Facade
All-glass facades are potentially expensive, even when they are designed with completely standardised, prefabricated units. Considerable problems are also caused by their connections to the load-bearing structure and in terms of fire protection. Other questions are posed by the internal comfort and thermal protection offered by such facades in summer. Where the glazing amounts to less than 70 per cent of the entire area, the selection of an appropriate glass type can obviate the need for additional external sun screening, which as a rule, accounts for roughly 30 per cent of the total facade costs (without having to accept an undesired change in the light spectrum).

Integrated building concepts
The greatest portion of the budget, however, is taken up by work below ground – a fact which is often, fortuitously, forgotten during early planning stages. A synergetic approach is usually helpful in this respect. If one builds on sites where the groundwater level is not much below the actual ground level, minimising the volume underground will obviously play a crucial role in achieving cost efficiency for the building as a whole. It may be possible to replace sub-floor spaces with facilities on the ground floor and thus avoid a second basement level. This, in turn, may help to avoid the creation of a waterproof tank. Similarly, if an underground garage is raised 1–1.2 metres out of the ground, mechanical ventilation can be avoided, the structure itself can be dimensioned more compactly, security within the spaces will be enhanced and a transition zone between public and private achieved.
The integrative concepts described here are experimental in nature and call for increased planning and powers of persuasion; and since no planner will want to guarantee the effectiveness of a cross-ventilation concept without trials, there are likely to be extensive carbon-dioxide tests and measurements after the building has been taken into operation. If these prove successful, however, there will be considerable savings in terms of investment and operating costs.

Standards
A further major potential for savings lies in the use of standardised elements. Buildings dating from the end of the 19th century, for example, drew on an established catalogue of parts. A large proportion of the elements used were prefabricated, such as facade units, doors, fireplaces, cast-iron railings, etc. The high standard of quality achieved in this era can, to a certain extent, be considered as resulting from tried and true, communicable and recognisable standards. They resulted in reasonable, calculable costs as well as durability. The latter was the outcome of a development process that made the elements appropriate for serial production. Paradoxically, it is much more difficult to convince people today of the advantages of such products. One reason for this is that the individual solutions with which we are familiar follow each other in ever more rapid cycles, before they can ever reach the design maturity that would make them suitable for serial production. We are quickly tired of the same things. Materials are used in rapid succession – wood, concrete, glass, synthetic combinations, and high-grade metals – all undergo various kinds of treatment. In an age in which surfaces play an important role, they are subject to quick changes, as in the world of fashion. On the other hand, especially in housing, certain elements like balconies, bathrooms and windows do not always need to be recreated. Individuality can be achieved through their use. Even if wet-rooms are locally prefabricated it is standardisation which, through tight specifications, simple construction logistics and repetitive processes, ensures cost savings.

3. Regulation

Untapped potential (especially in comparison with some European neighbours) makes one recognise all those phenomena that could be grouped under the heading "regulation". Construction in Germany is comprehensively and tightly controlled by various codes and standards, and there are many redundancies and overlaps in this respect. For example, it is hard to understand why planners and project managers are controlled by so many different fire protection agencies and construction sites by so many different accident and employee associations. A tight network of regulations and all their ramifications impose burdens on planning and construction alike. A sensible structural deregulation of planning and site processes would release the great potential that exists both here and abroad and lend new meaning to the idea of cost awareness in building.

Conclusion

The essence of cost awareness in building is presented in the aspects discussed above. There is sufficient potential here to achieve a recognisable reduction in building costs, and reason enough to join our neighbouring countries in new approaches of cost-controlled construction. Decisive is, of course, that quality must not be allowed to suffer. In housing construction, for example, rather than thermal insulating composite panels, high quality double-layer facades could be considered, as are applied in the Netherlands. Critical consideration should be made in the area of deregulation of quality standards which are still comparatively high in Germany; one alternative is the development of different standards for different situations. This would lead to market opportunities for new products with varying prices.

1 Primary School Theresienhöhe in Munich
2 Otto-Steidle Atelier in Munich

Student Hall of Residence in Amsterdam

Architects: Claus en Kaan Architecten, Amsterdam/Rotterdam

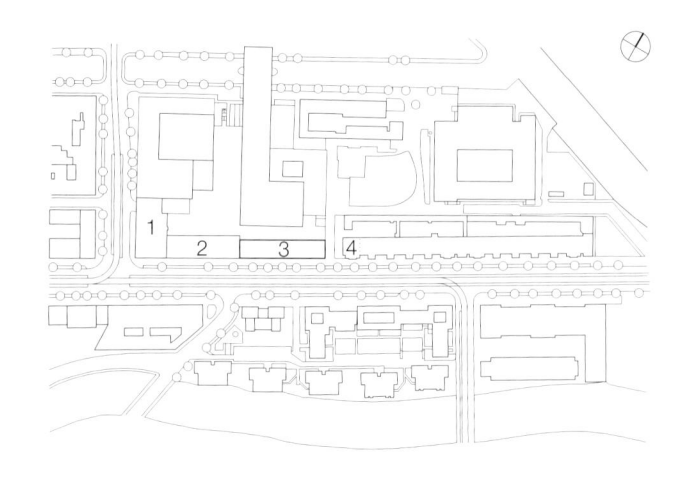

Room sizes based on minimum recommendations
Facade language with off-set windows

This hall of residence is part of a new development intended to recreate the urban environment which existed prior to various building schemes from the 1960s. The master plan, by Pi de Bruyn from de Architecten Cie, reinstates the typical Amsterdam street frontage, and contains three discrete stages. The corner building was designed by de Bruyn and the central section by VMX architects. The third building stage, by Claus en Kaan Architecten, is an urban repair project; although there is little evidence of sentimentality or nostalgia, and visible association with the historical environment has been avoided.

Facade treatment

The irregular arrangement of the windows is a subtle, contemporary version of the decorative window and door language of the neighbouring constructions. The ground floor is given over to shops, while above are 61 student apartments. A cross-wall construction of reinforced concrete forms the building's frame and the walls fronting the street are of a lightweight construction clad with brick. Flush-set glass and stainless steel elements of varying dimensions inject rhythm and contrast into the ground floor facade. The upper level windows are deeply set into the reveals, and present an off-set pattern. Alone the stairwells and entrance area break out of the rigid cross-wall construction system and are separated from one another by a curved wall and an arched suspended ceiling. These colourful elements and the curtains in the apartments create sharp contrast to the black coating of the exposed external brickwork.

Economical living

The residences were to be rented out as economically as possible. The dimensions of the individual spaces correlate exactly with minimum building requirements. Each apartment benefits from a kitchenette, obviating the need for communal cooking and dining areas. Communication between the students takes place in the covered walkways which interconnect the various apartments and storage spaces. It was also considered unnecessary to provide a basement level.

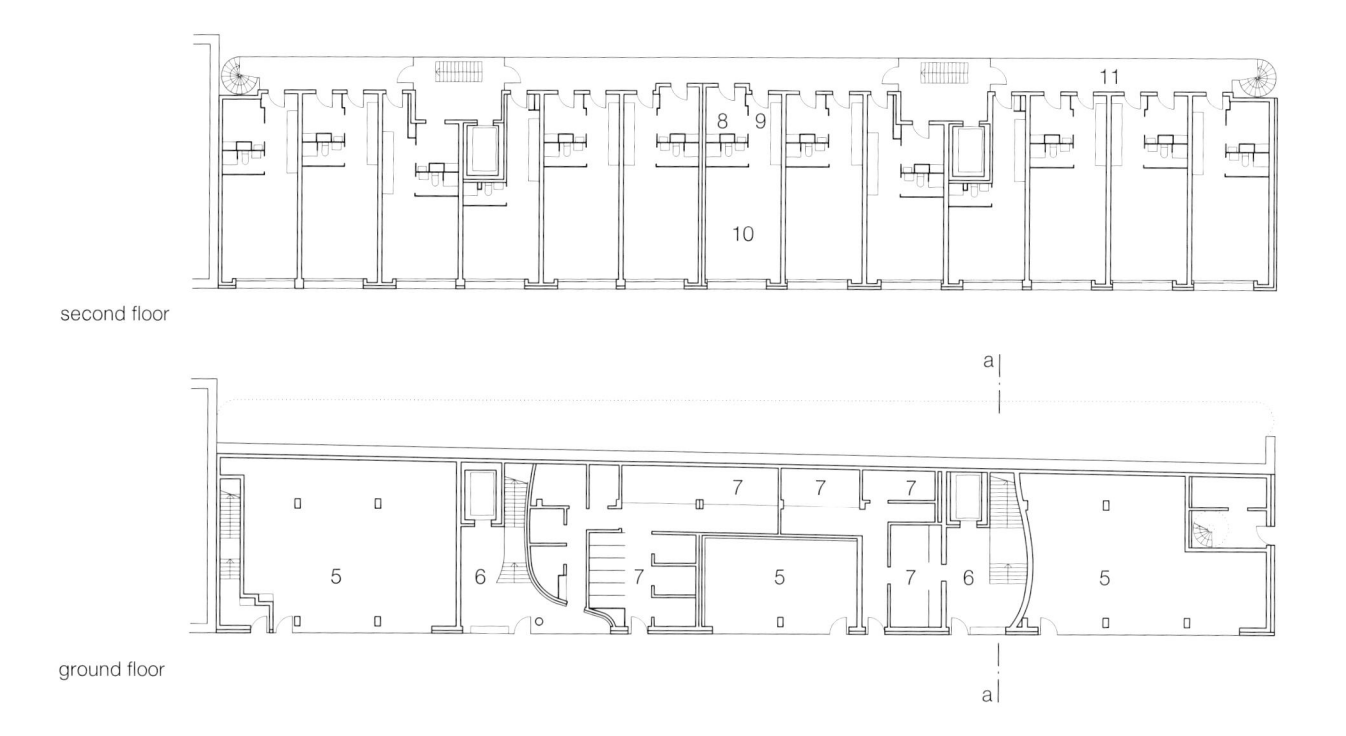

second floor

ground floor

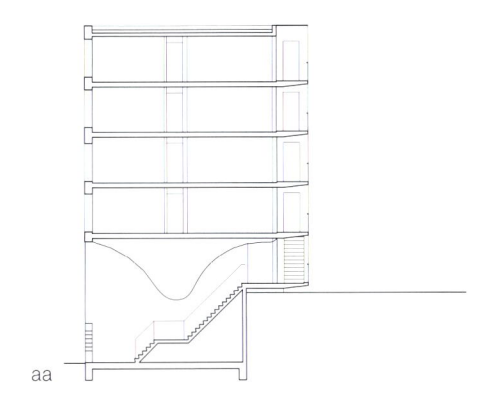

section · floor plans   scale 1:400
site plan   scale 1:5000

1 building stage by de Architecten Cie (Pi de Bruyn)
2 building stage by VMX architects
3 building stage by Claus en Kaan Architecten
4 building stage by Claus en Kaan Architecten (not realised)
5 shop
6 entrance hall
7 general and bicycle storage
8 storage
9 kitchen
10 student room
11 covered walkway

aa

Project details:
Usage:	hall of residence retail (ground floor)
Residential units:	61 apartments
Construction:	reinforced concrete/ exposed brickwork
Total construction cost:	3.6 million Euros
Cost per m² area:	986 Euros
Total internal volume:	9,000 m³
Total built area:	3,650 m²
Internal ceiling height:	2.44 m
Date of construction:	2002

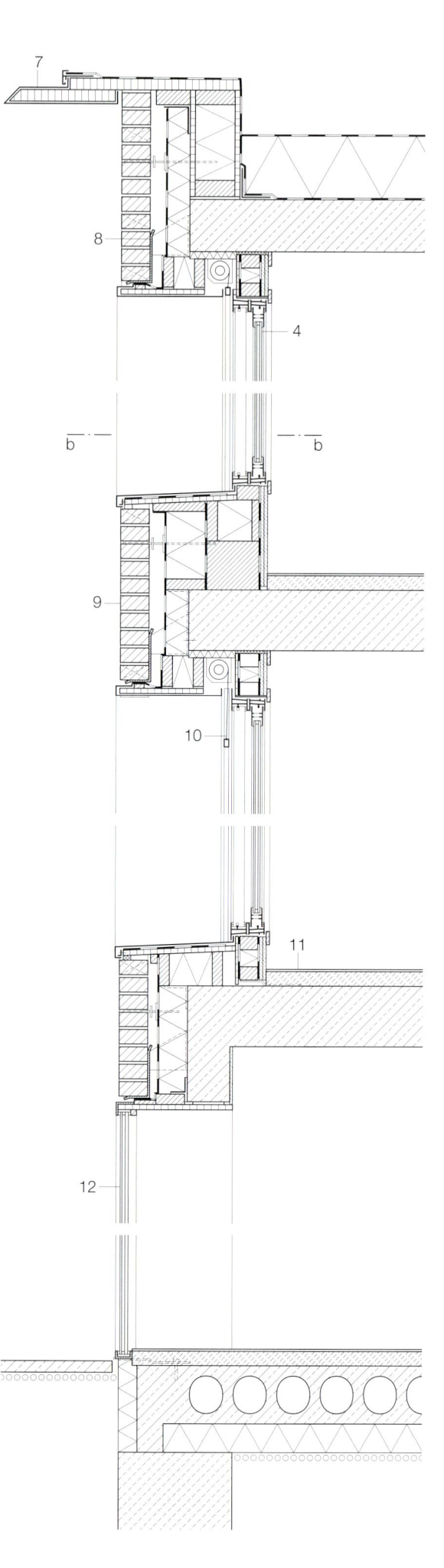

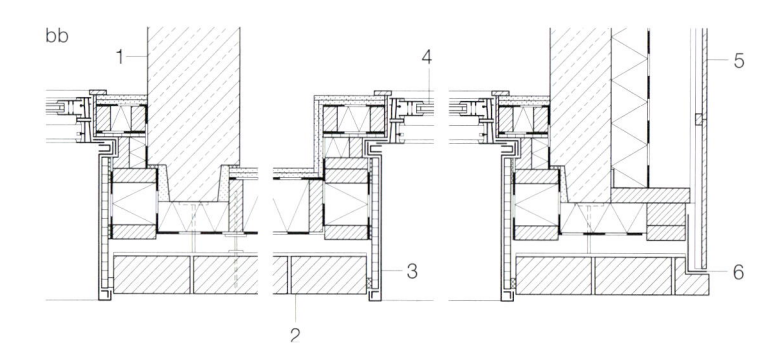

horizontal section
vertical section
scale 1:20

1 250 mm reinforced concrete
2 100 mm masonry, coated black
   60 mm ventilation cavity
   permeable membrane
   140 mm mineral wool
   vapour barrier
   2× 12.5 mm plasterboard
3 anodised aluminium sheet
   18 mm veneered plywood
4 aluminium sash window with
   sound-insulating double glazing
   8 mm float + 9 mm cavity
   + 2× 6 mm toughened glass
5 4 mm fibre-cement shingles
   38 × 19 mm battens
   38 × 15 mm counterbattens
   117 mm ventilation cavity
   permeable membrane
   100 mm mineral wool
   170 mm reinforced concrete
6 60 × 150 × 4 mm sharp-cornered
   steel angle
   with baked enamel finish
7 zinc sheet bent to shape
   40 mm veneered plywood
8 welded steel sheet bracket
   anchor
9 100 mm coated masonry
   60 mm ventilation cavity
   permeable membrane
   80 mm mineral wool
10 roller awning
11 2.5 mm linoleum floor covering
   50 mm cement-based monolithic
   screed
   210 mm reinforced concrete
   floor slab
12 double glazing; 8 mm float glass
   + 15 mm cavity
   + 10 mm laminated safety glass

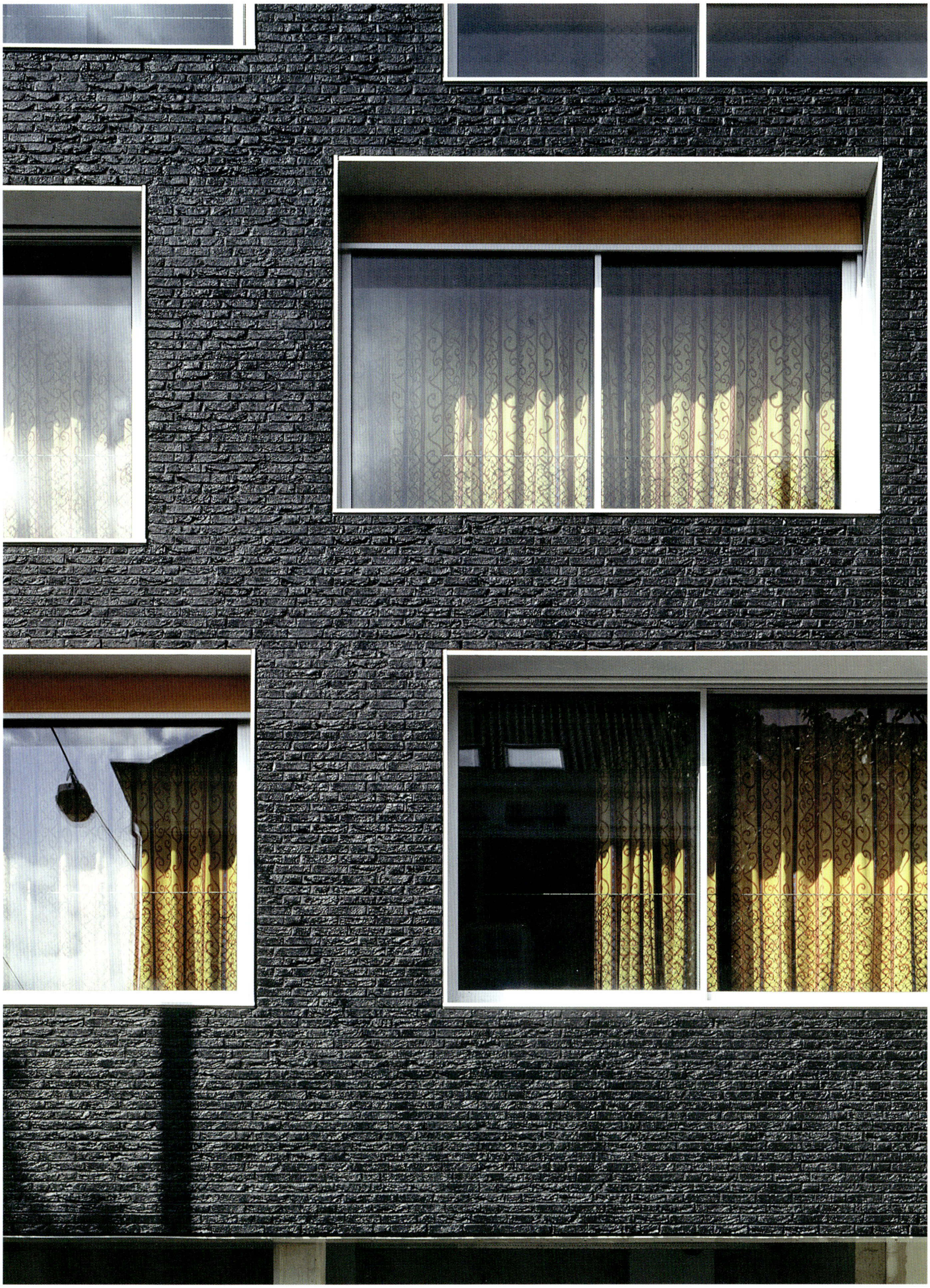

Youth Camp in Passail

Architects: Holzbox Tyrol, Innsbruck

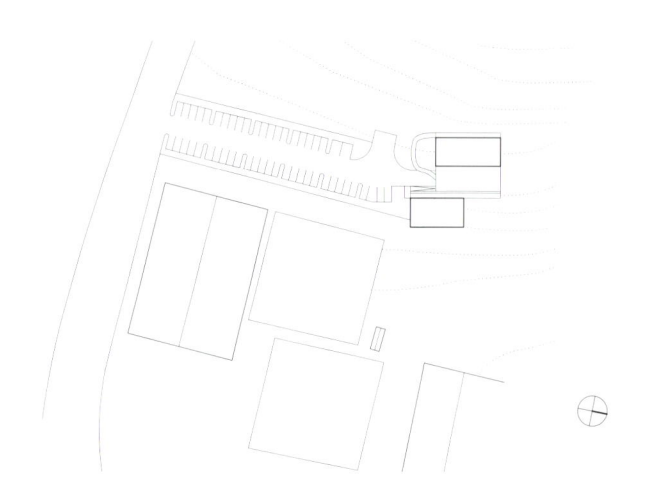

**Modular system independent of location
Cost efficiency due to high level of pre-fabrication**

"Multifunctional Camp Modules" was the title of the competition whereby the Austrian state of Styria sought alternative concepts to traditional youth hostels; normally one-off schemes built independently of one another. This system, based upon sustainable modular construction techniques and which can be transferred to almost any desirable location, is to provide a unified, corporate image for the future-oriented tourism industry of the region.

Wooden containers as apartments

The winning project consists of 9.8-metre-long wooden containers, with nominal interior dimensions, which reduce the necessary floor space to a minimum. Three different module sizes are available for various user groups, all with internal ceiling heights of 2.5 metres: a two-metre-wide container for one carer, a three-meter-wide youth module, and an apartment module of 40 m^2 for up to eight beds. Originally the architects intended to have the individual modules prefabricated as solid timber elements and transported to the site by road. Because the access to this site was too narrow for an oversized load, they made an exception with this first youth camp in Passail, specifying instead a sandwich-panel construction with prefabricated walls, floors and ceilings. The facades, interiors and the custom-designed furniture are also created of modular elements. By simply relocating a sliding door and partition wall, it is possible to adjust the number and types of beds; a double bed effortlessly becomes two bunk beds. The entire planning and construction process of the ten apartments, which contain a total of fifty-eight beds, was completed in four months. The living containers rest on bearers and concrete footings and are divided between two buildings, which combine with an existing tree to create an open terrace area for outdoor recreation. This activity space extends underneath the cantilever of the upper building toward the communal room on the ground floor which provides an alternative for rainy days.

Project details:
Usage: youth hostel
Units: ten apartments with 58 beds
Construction: timber sandwich-panel system
Internal ceiling height: 2.5 m
Total construction cost: 708,000 Euros
Cost per m² usable area: 1,500 Euros, approx.
Total internal volume: 1,800 m³
Total built area: 550 m²
Total usable area: 470 m²
Site area: 5,535 m²
Built site area: 395 m²
Date of construction: 2004
Construction period: May–July 2004

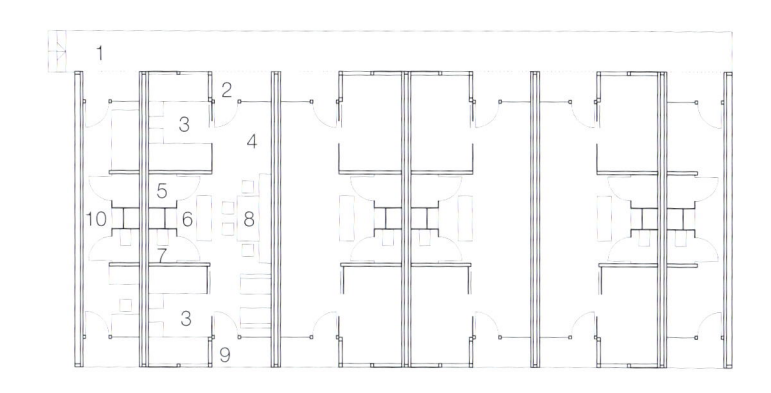

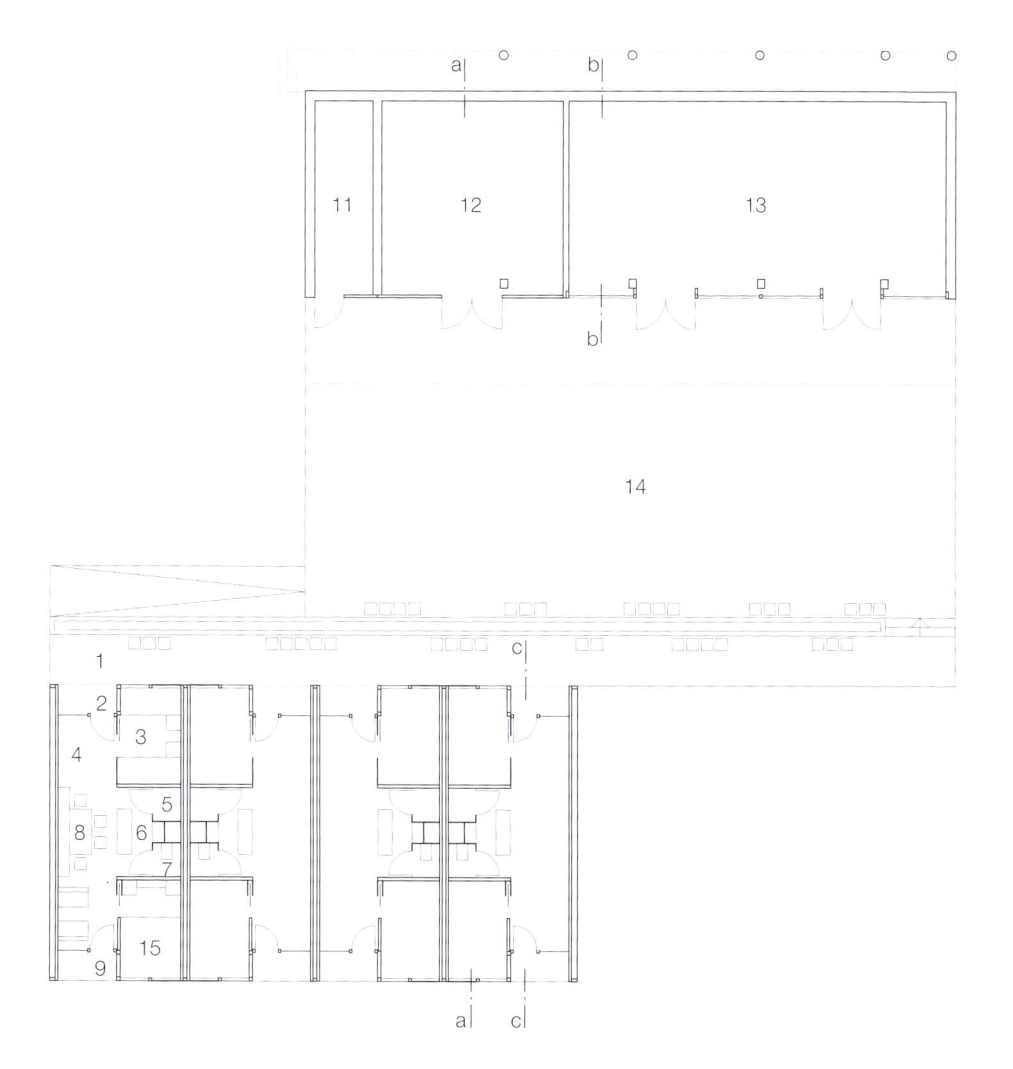

site plan
scale 1:2500

floor plans
upper floor
ground floor
scale 1:250

1 entry bridge
2 entrance
3 children's berths
4 cloakroom
5 shower
6 hand basin/kitchenette
7 WC
8 dining
9 balcony
10 apartment for carer
11 technical services
12 storage
13 communal room
14 courtyard
15 parents' berths

97

bb

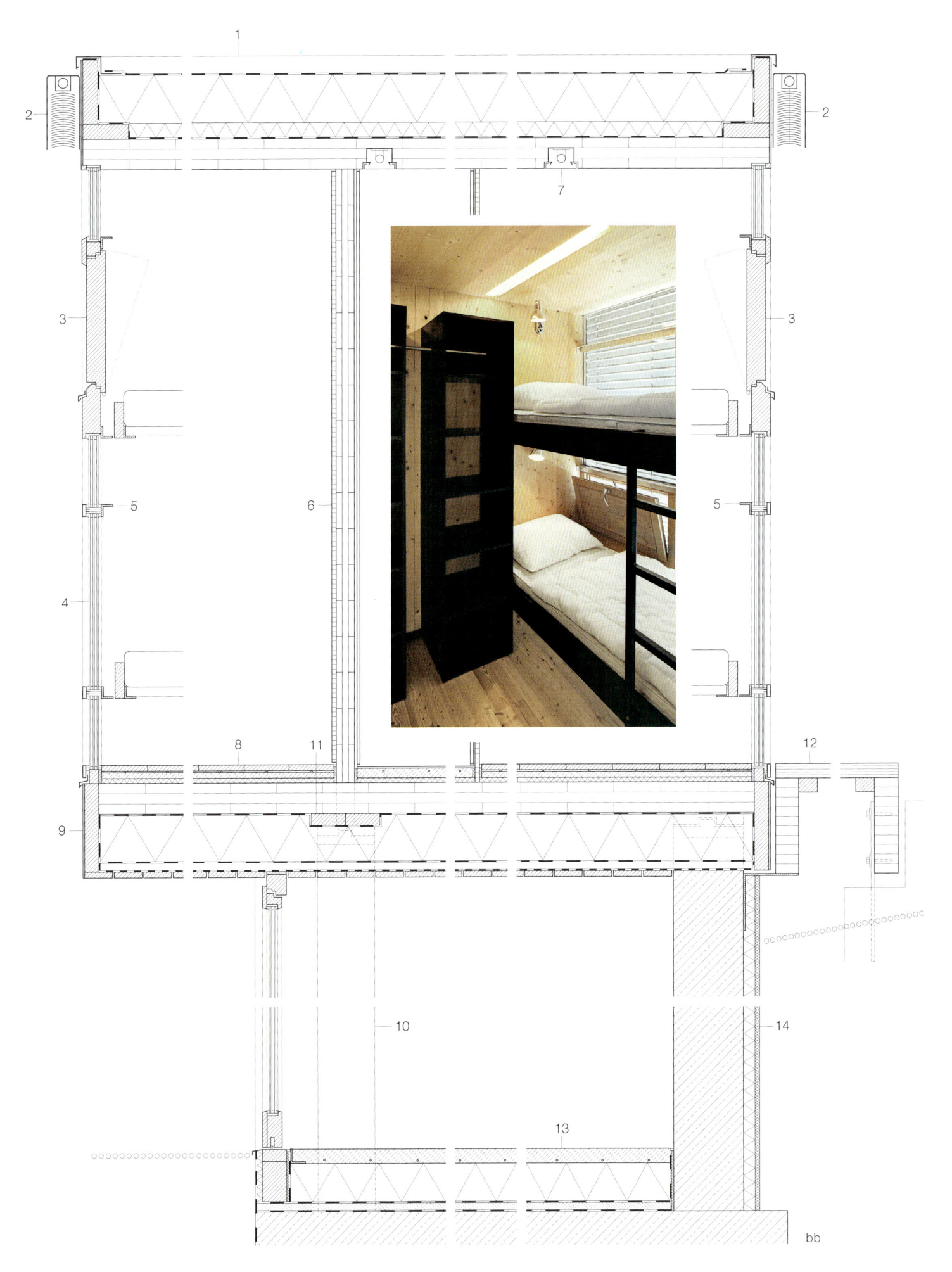

98

section   scale 1:250
vertical section   scale 1:20

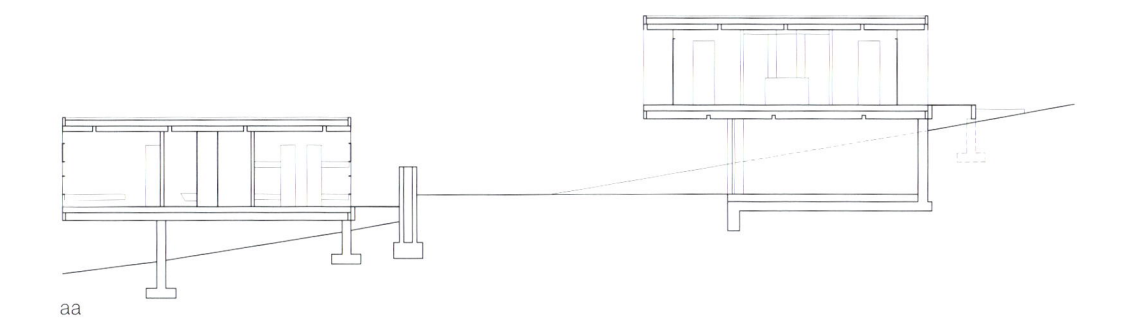

aa

1   4273 × 9800 mm roof element:
    1.5 mm EPDM membrane sealant
    150–200 mm insulation panels to falls
    30 mm insulation
    vapour barrier
    128 mm solid fir panel glued crosswise
2   1 mm aluminium casing sun protection
3   80 mm solid larch ventilation element
4   glazing: 5 mm + 12 mm cavity + 5 mm
    + 12 mm cavity + 5 mm
5   50 × 50 × 5 mm steel angle facade transom
6   partition wall between berths and shower:
    19 mm medium-density-fibreboard
    79 mm solid fir panel

    13 mm laminated board, green
7   recessed ceiling light fixture
8   floor construction apartment:
    24 × 120 mm larch boarding,
    underlay mat
    electric under-floor heating on
    20 mm dry-screed substructure
    20 mm sound-impact insulation
    128 mm solid fir panel, vapour barrier
    200 mm rockwool, wind-proofing
    30 mm battens with ventilation cavity
    black felt, watertight under driving rain
    24 × 120 mm larch boarding
9   5 mm laminated board

    60 × 200 mm timber header
10 250 × 250 mm reinforced concrete column
11 2, 50 × 150 mm larch reinforcement struts
12 60 × 120 mm larch entry bridge
13 floor construction basement:
    60 mm heating screed with treated surface
    160 mm expanded polystyrene
    thermal insulation
    PE-membrane, 30 mm filling, PE-membrane
    300 mm reinforced concrete
14 wall construction:
    drainage panel
    50 mm extruded polystyrene insulation
    300 mm watertight reinforced concrete

99

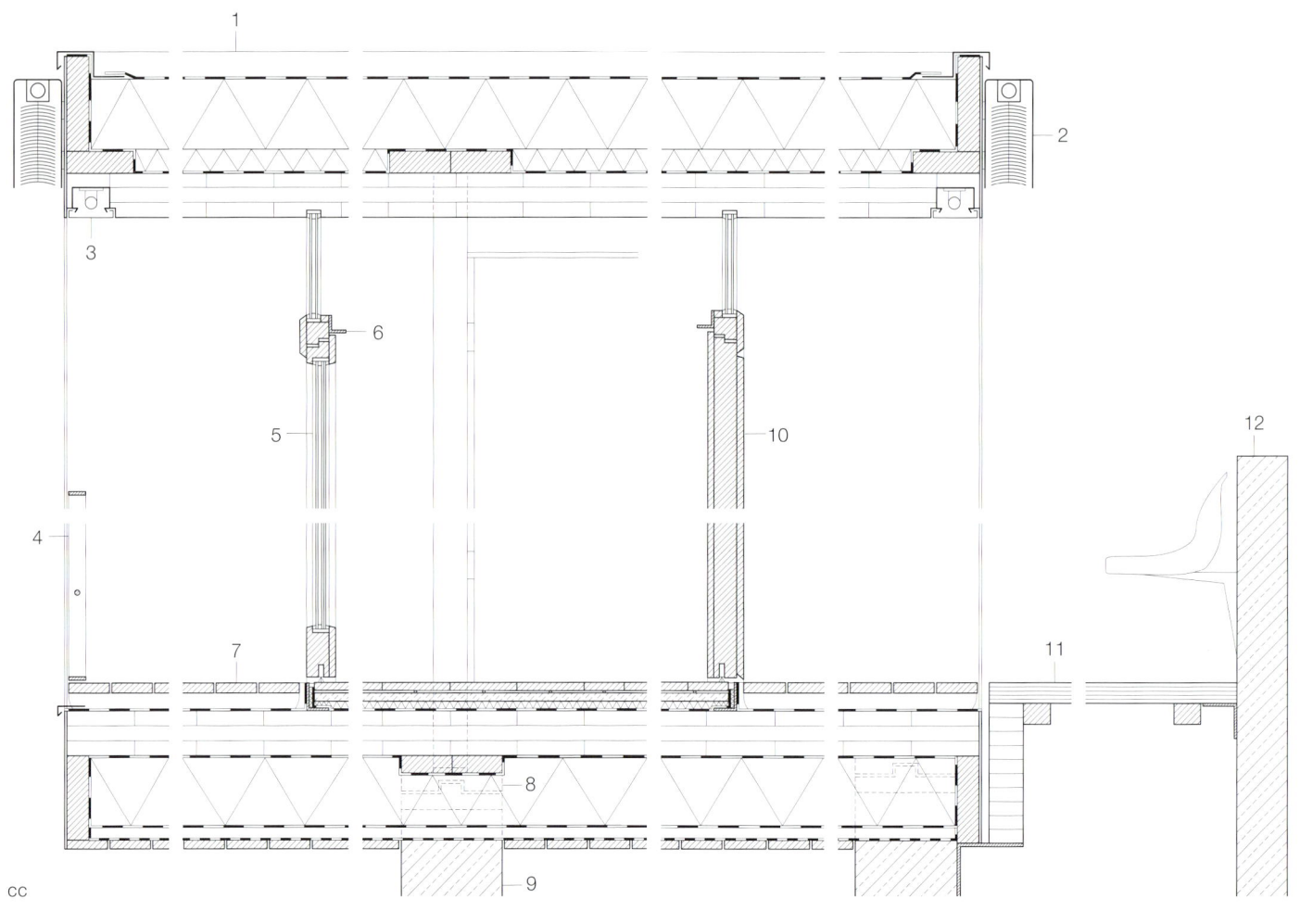

cc

100

vertical section
horizontal section
scale 1:20

1. 4273 x 9800 mm roof element:
   1.5 mm EPDM-membrane sealant
   150 - 200 mm insulation panels to falls
   30 mm insulation, vapour barrier
   128 mm solid fir panel glued crosswise
2. 1 mm aluminium casing sun protection
3. recessed ceiling light fixture
4. 50 × 10 mm steel flat railing
5. balcony door glazing:
   5 mm + 12 mm cavity + 5 mm
6. 50 × 50 × 5 mm steel angle facade transom
7. floor construction balcony:
   24 × 120 mm larch planks, battens
   1.5 mm EPDM sealant, 128 mm solid fir panel
   vapour barrier
   200 mm rockwool, wind-proofing
   30 mm battens with ventilation cavity,
   black felt
   24 × 120 mm larch boarding
8. 250 × 250 × 20 mm elastomer bearer
9. 300 mm reinforced concrete
10. 40 mm solid larch entrance door
11. entry bridge:
    24 × 120 larch boarding on
    400 × 50 mm laminated timber bearers
12. 150 mm reinforced concrete plant box
13. partition wall between shower and
    sleeping berth:
    13 mm laminated board, 6 mm sealing strip
    79 mm solid fir panel
    19 mm black medium-density-fibreboard
14. bathroom door:
    13 mm green laminated board
15. bathroom cabinet with mirror
16. party wall:
    79 mm solid fir panel
    2× 60 mm acoustic insulation
    79 mm solid fir panel
17. sliding door/wall:
    38 mm medium-density-fibreboard,
    removable for bed re-arrangement
18. triple glazing: 5 mm + 12 mm cavity + 5 mm
    + 12 mm cavity + 5 mm U = 0.7 W/m^{2}K
19. 80 mm solid larch ventilation element
20. 19 mm medium-density-fibreboard,
    vapour barrier
    100 mm thermal insulation, wind-proofing
    5 mm laminated board
21. external wall element:
    24 × 120 mm larch boarding
    permeable black felt
    36 mm battens with ventilation cavity
    wind-proofing, 140 mm rockwool
    79 mm solid fir panel

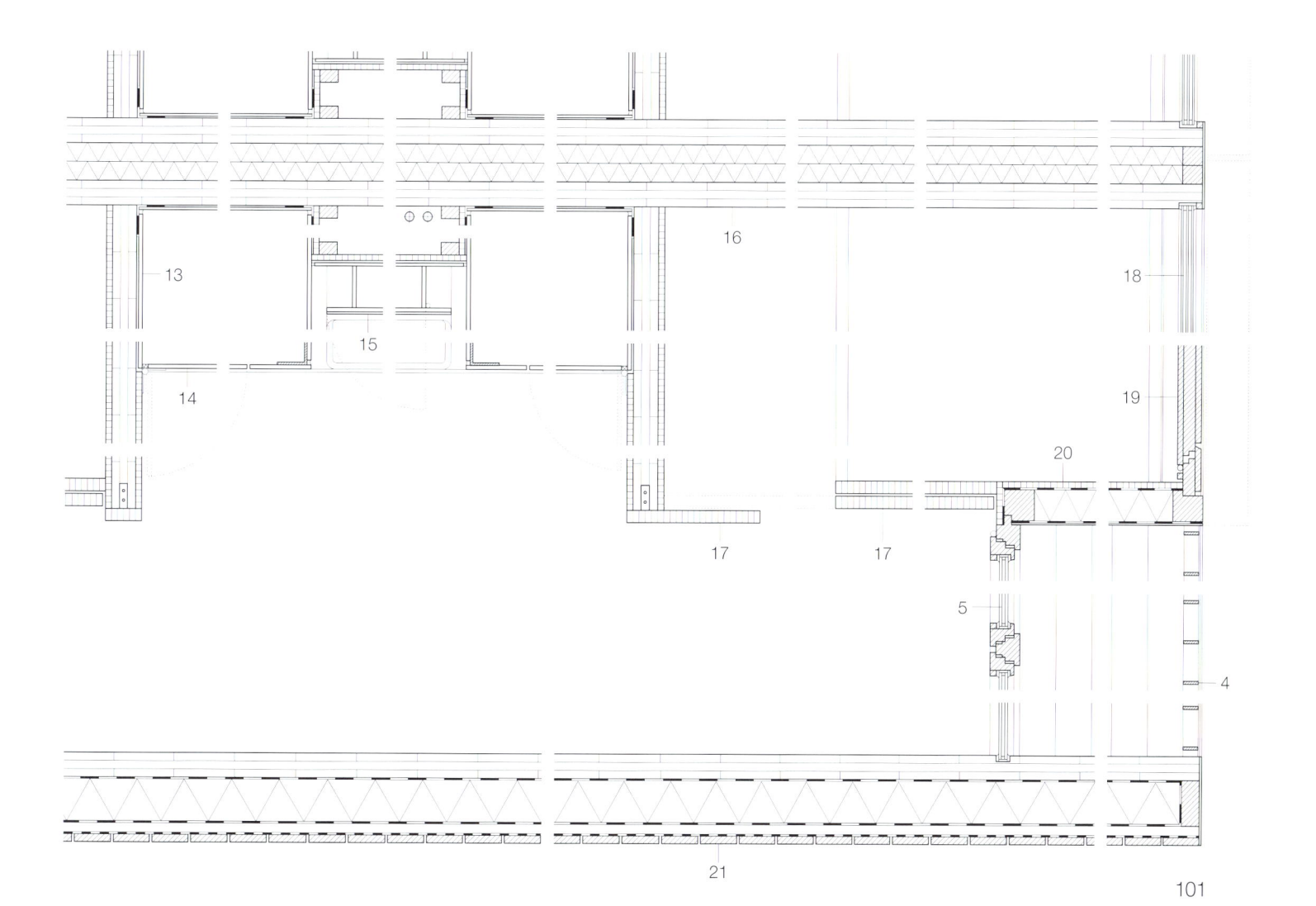

Hotel in Groningen

Architects: Foreign Office Architects, London

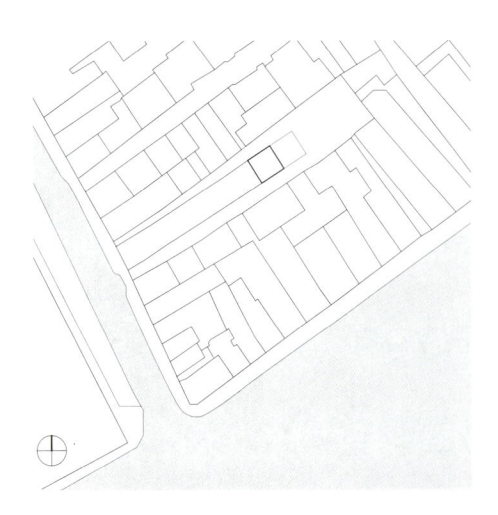

Architectural event under the guidance of Toyo Ito as starting point

Going by the name of "Blue Moon", an architectural event in the Netherlands took place under the guidance of Toyo Ito in 2001. In addition to the temporary buildings constructed at various locations, some long-term structures were also created; for example the Aparthotel in the "Schuitenwerks" district of Groningen. The appearance of this area is heavily influenced by trade; with small docks, warehouses and guest houses for travellers and traders shaping the urban image.

External appearance changes with use

An in-fill construction in a small square, this striking structure resembles a plain storage building rather than a hotel. As the level of activity increases over the course of the day, however, the appearance of the building changes accordingly. The various shutters and doors begin to open, in the bar in the two lower levels and the hotel apartments above, physically revealing the functions, and transforming the appearance of the building. The outer skin of the structure also undergoes a change at night. The corrugated aluminium facade which appears opaque by day is, in fact, finely perforated; the internal life of the building shimmers through at night when the interior lights are turned on.

Attractive meeting place on the roof terrace

The construction of a hotel with café has revitalized the small square and created an attractive meeting place for the district. From the roof terrace, guests benefit from views over the surrounding old town centre.

Project details:
Usage:	Hotel with two apartments of 37.7 m² each and a café
Construction:	steel
Internal ceiling height:	2.6–3.6 m
Total construction cost:	450 000 €
Cost per m² total built area:	2,142 €
Total internal volume:	598 m³
Total built area:	210 m²
Date of completion:	August 2001

site plan
scale 1:2000
floor plans
scale 1:200

1 café
2 access to apartments
3 apartment
4 bathroom
5 roof terrace
6 services

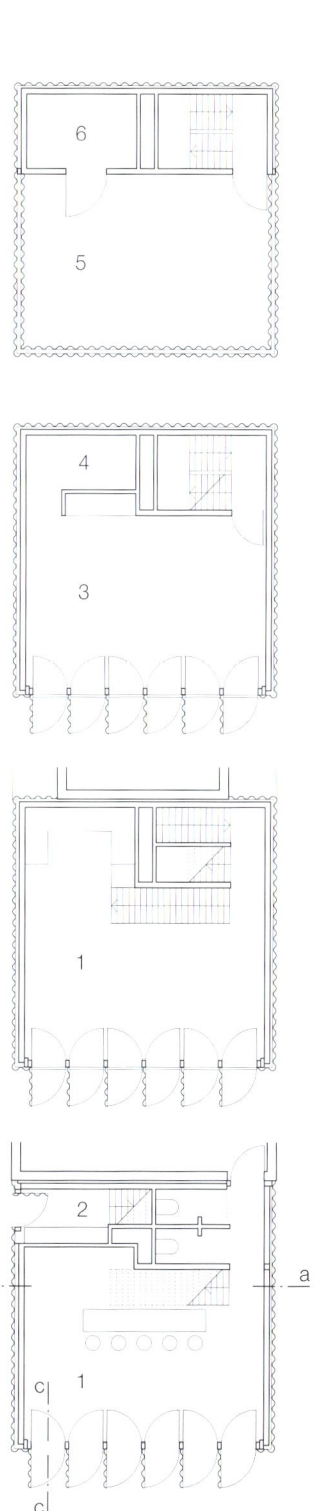

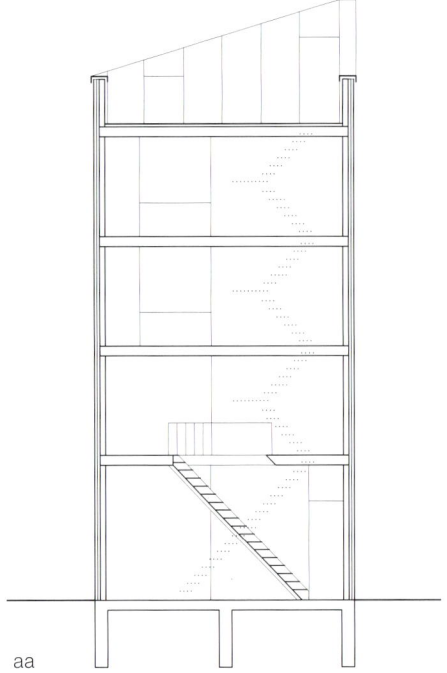

section   scale 1:200
horizontal section • vertical section
scale 1:10

1   wall construction:
    40 × 160 mm aluminium-zinc
    corrugated sheeting
    30 × 20 mm steel SHS supporting structure
    90 mm thermal insulation
    150 mm sand-lime brickwork
2   140 × 50 mm aluminium post
3   aluminium casement with double glazing:
    8 + 8 mm laminated safety glass
4   30 × 30 × 3 mm steel T-section
5   65 × 30 mm steel angle
6   70 × 50 mm steel RHS
7   160 × 40 mm perforated aluminium-zinc
    corrugated sheeting
8   60 × 60 × 3 mm steel T-section
9   50 × 50 mm steel SHS
10 roof construction:
    30 mm concrete paving slabs
    bituminous sealing layer
    60 mm thermal insulation
    vapour barrier
    200 mm reinforced concrete roof slab
11 steel bracket for fixing hinge
12 30 × 50 mm steel SHS safety rail
13 25 mm composite wood board
14 15 mm thermal insulation
15 18 mm concrete pavers
16 prefabricated concrete plinth

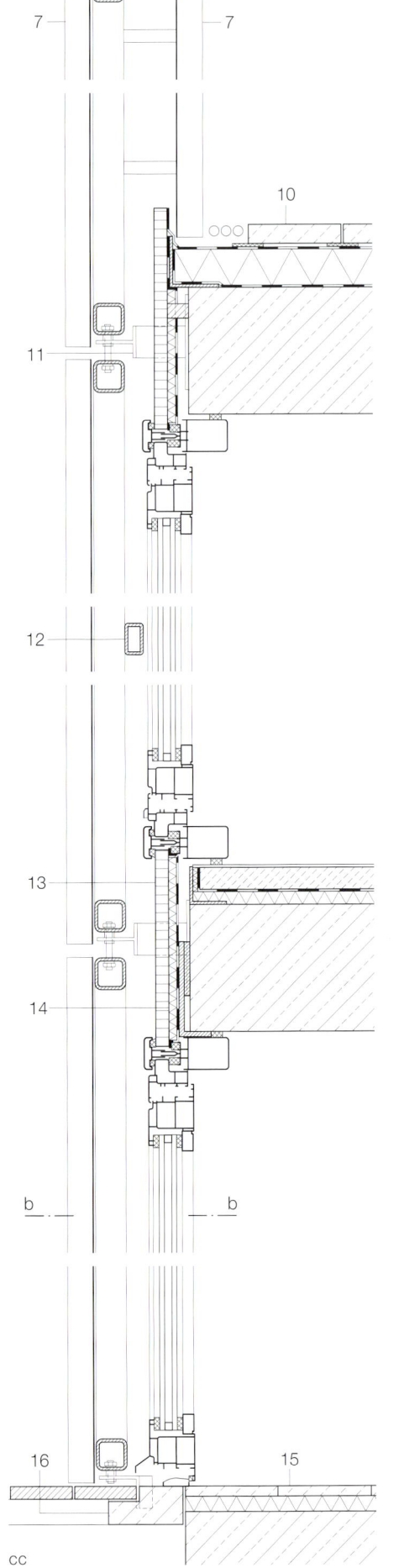

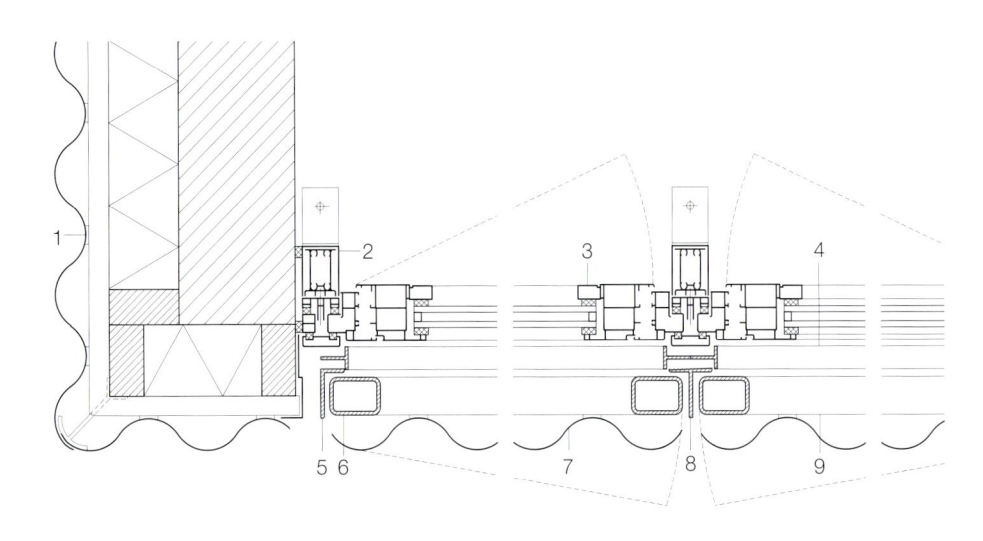

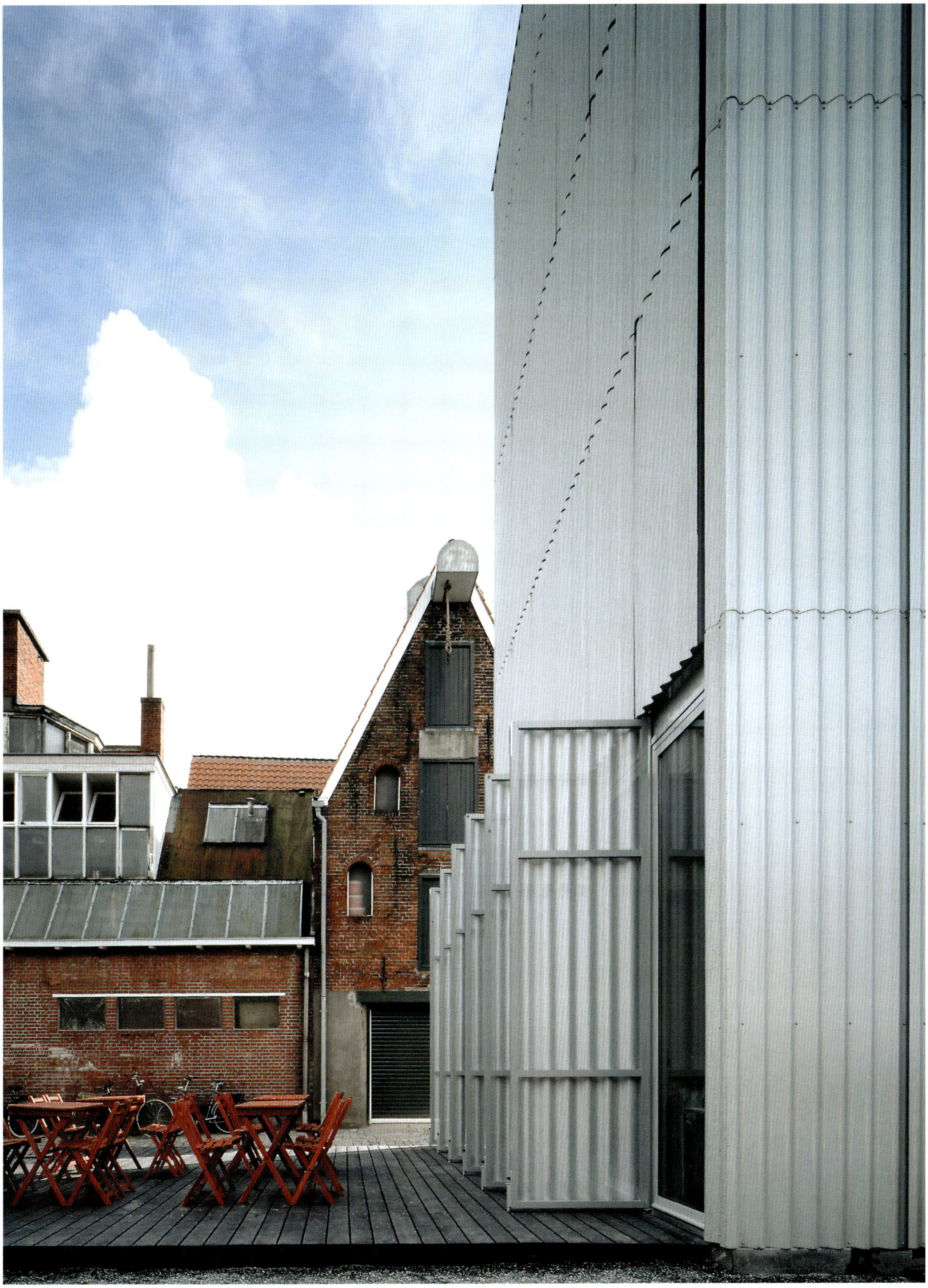

Cultural Centre in Munich

Architects: Ingrid Amann Architekten, Munich

**High level of prefabrication
Precast concrete construction techniques
Implementation of standard industrial products**

Reflecting its surroundings like still life images of the suburbs, the windows of the Trudering Cultural Centre reveal the neighbouring single-family dwellings, used car yards and supermarkets. A heavily used axial thoroughfare of Munich, the Wasserburger Landstrasse generates an immense level of traffic noise, none of which can be heard from within the centre. Situated immediately adjacent to this arterial road, the centre is a visible cultural landmark in an otherwise commercial urban area.

Large-scale elements creating open spaces
The 45-metre-long cubic structure presents a smooth exterior to the roadway. To the South, however, the cantilevered upper level creates a protected roof over the terrace. The two sides of the building are connected by an internal walkway with a width of one construction axis which slices through the structure; to the North it acts as a forecourt while internally it widens and becomes a two-storey foyer crossed by a variety of upper galleries. The various communication zones of the building interconnect to create continuous access through the foyer and the galleries of the main hall. The permeability and complexity of the spatial sequences can best be appreciated with the doors of the hall thrown open. Views are achieved through the foyer, over the terrace and into the landscape beyond.

Construction costs were reduced by early and intense detailed planning, by the use of prefabricated elements and by dispensing with a basement. The load-bearing structure consists of a concrete skeletal framework laid out in five bays on an 8.9-metre grid. The 27-metre-long and up to 1.5-metre-deep north-south beams span a distance of 18 m over the hall and cantilever out by 6 m in the area of the group rooms on the southern side. Suspended columns are concealed within the concrete sandwich elements on this facade. The load-bearing roof over the group rooms consists of sloping trapezoidal-section metal sheeting while the beams in the centre of the hall were raised in order to produce the necessary fall for the flat roof. Heavy graded concrete or expensive graded thermal insulation panels could thereby be dispensed with. The roof over the northern rooms was constructed of pre-stressed hollow concrete slabs for sound-insulation purposes; the underside of which is clad only in the main hall where the ventilation system is concealed above the suspended ceiling. The exposed, perforated metal sheeting acts as both a load-bearing structural element and acoustic ceiling; a suspended ceiling was not considered necessary in this area. Precast concrete filigree elements were specified for the intermediate floor slabs, thus avoiding the use of formwork. Standard industrial products provided the elements for the economical dome and strip-shaped skylights. By being set flush into the construction the appearance of the skylights was greatly enhanced. In order to do away with as many complicated finishing procedures and secondary constructions like double walling, steel consoles or metal cover plates, it was considered necessary for the prefabricated concrete panels to be individually detailed in regard to the connections. The high level of prefabrication allowed this procedure to be carried out without increases in manufacturing costs. The wall units were designed to a maximum size to reduce the number of joints and fixings.

Permanent colour concept
The sharply edged concrete slabs were pigmented and polished in the factory creating highly reflective surfaces. The reddish tones of the facades relate to the colouration of the surrounding tiled roofs while the warm grey of the gallery balustrade panels and the rear wall of the main hall form a backdrop like a stage set. Initial scepticism on the part of the population about the use of industrial precast concrete for a cultural centre has given way to a broad degree of acceptance. This is attributable to the bright, ample spaces, the quality of the facade surfaces and the careful yet robust details.

Project details
Usage: Local cultural centre
Construction: reinforced concrete
Internal ceiling height: 3.2 m
Construction cost: 1.8 mill. Euros
Cost per m² usable area: 1,100 Euros
Total internal volume: 9,478 m³
Total floor area: 1,900 m²
Usable area: 1,705 m²
Site area: 26,430 m²
Cost of construction: 1.8 million Euros
Cost per area: 1,100 Euro/m²
Construction period: 15 months

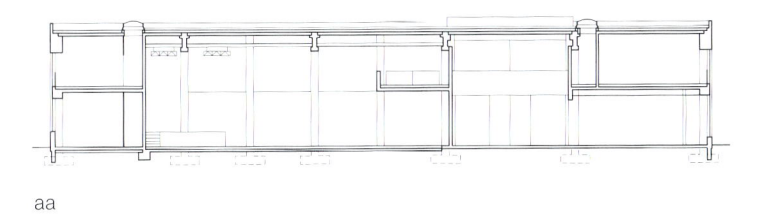

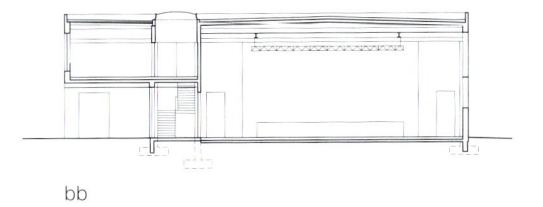

site plan
scale 1:2500
floor plans · sections
scale 1:500

1 festival ground
2 entrance
3 foyer
4 great hall
5 performers
6 services
7 chair store
8 terrace
9 group room
10 void
11 balcony in hall
12 gallery

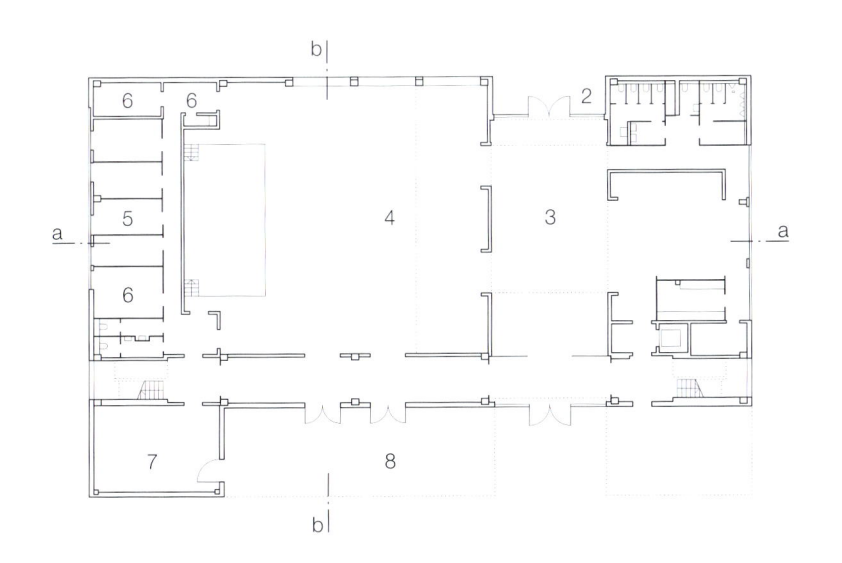

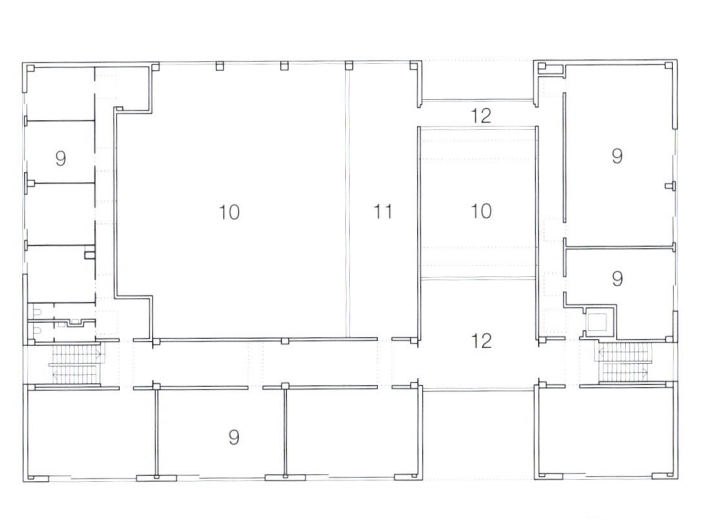

aa

bb

bb

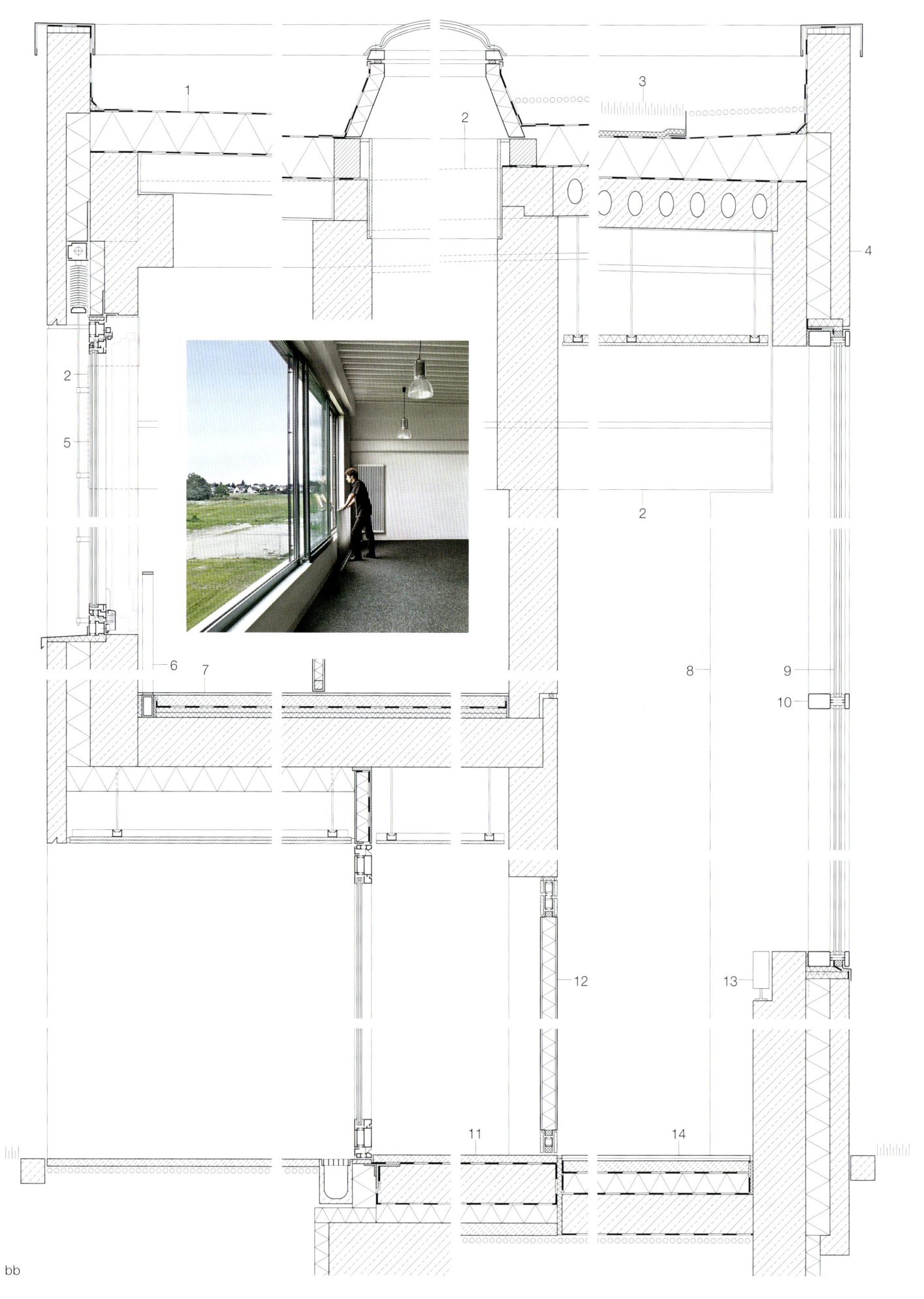

108

vertical section
horizontal section first floor
scale 1:20

1. 0.2 mm impermeable membrane
   160 mm insulation, vapour barrier
   160 mm trapezoidal-section metal sheeting with acoustic perforations
2. 27000 × 500 mm reinforced concrete beam, top graded to falls
3. 80 mm planted layer with sedum
   20 mm drainage layer, separating layer
   160 mm insulation, vapour barrier
   8300 × 1200 × 200 mm hollow tensioned concrete
   50 mm insulated acoustic ceiling
   12.5 mm plasterboard
4. sandwich element:
   80 mm reinforced concrete,
   pigmented and polished with sharp arrises
   100 mm insulation
   140 mm reinforced concrete
5. aluminium sliding/bottom-hung casement
6. 45 × 5 mm flat steel safety barrier
7. 10 mm needle felt, 50 mm screed
   separation layer
   2× 22 mm impact-sound insulation
   220 mm reinforced concrete filigree slab
   160 mm insulation
   2× 12.5 mm moisture resistant plasterboard
8. 400 × 500 mm reinforced concrete column
9. sound-insulating glazing (R'w = 41 dB)
10. 60 × 100 mm steel rod RHS post-and-rail construction
11. 30 mm asphalt, glass-fibre mat,
    160 mm reinforced concrete slab
    bituminous sheeting
    80 mm insulation
    50 mm foundation course
    200 mm gravel
12. 30 min. fire-resistant door with retractable bottom seal
13. convector
14. 22 mm fumed oak parquet
    45 mm anhydrite screed
    Polyethylene membrane
    160 mm reinforced concrete slab
    200 mm gravel
15. 400 × 200 mm reinforced concrete suspended column
16. aluminium facade ventilating element
17. glazing to group room:
    6 mm toughened glass + 16 mm cavity + 8 mm laminated safety glass (R'w = 34 dB)
18. double glazing to foyer (8 + 16 + 6 mm)
19. 3 mm aluminium sheeting
    180 mm insulation, vapour barrier

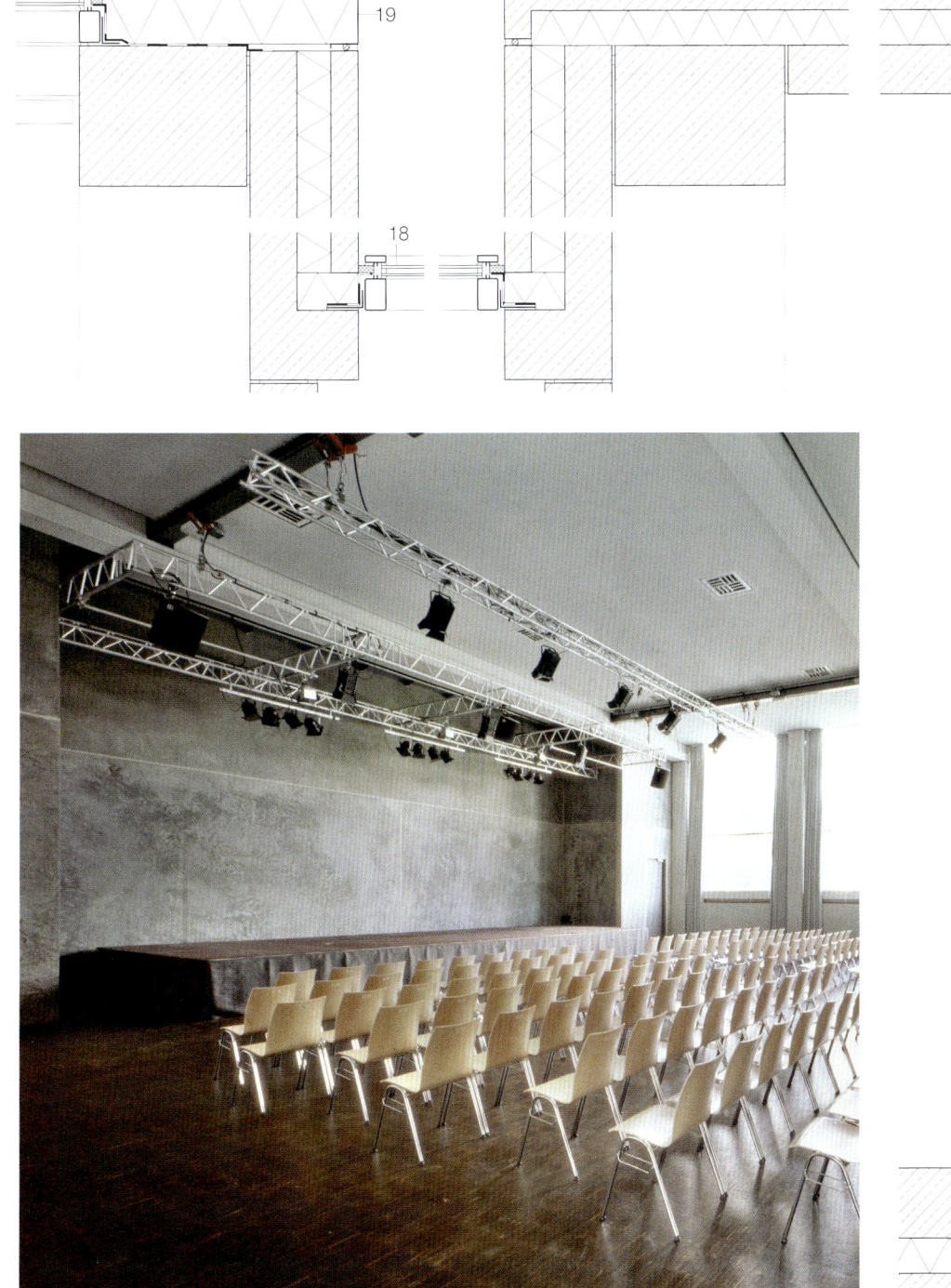

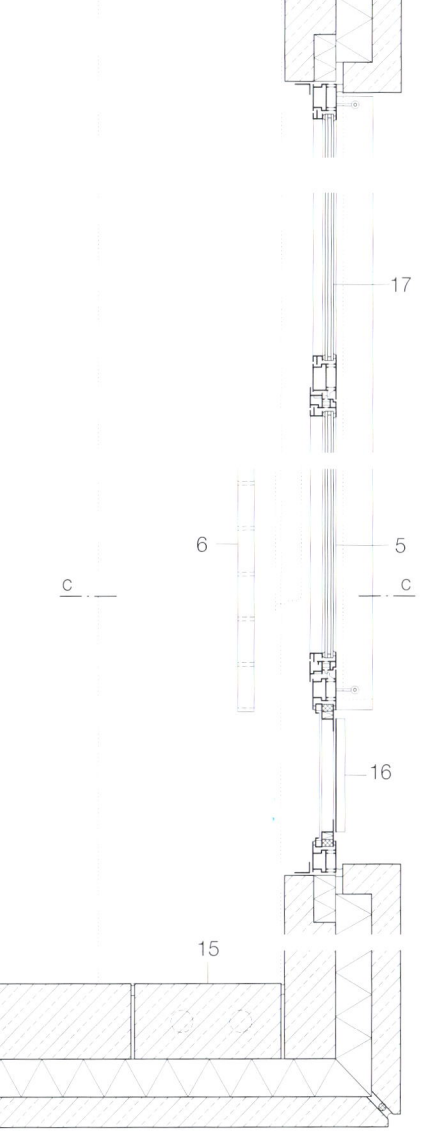

109

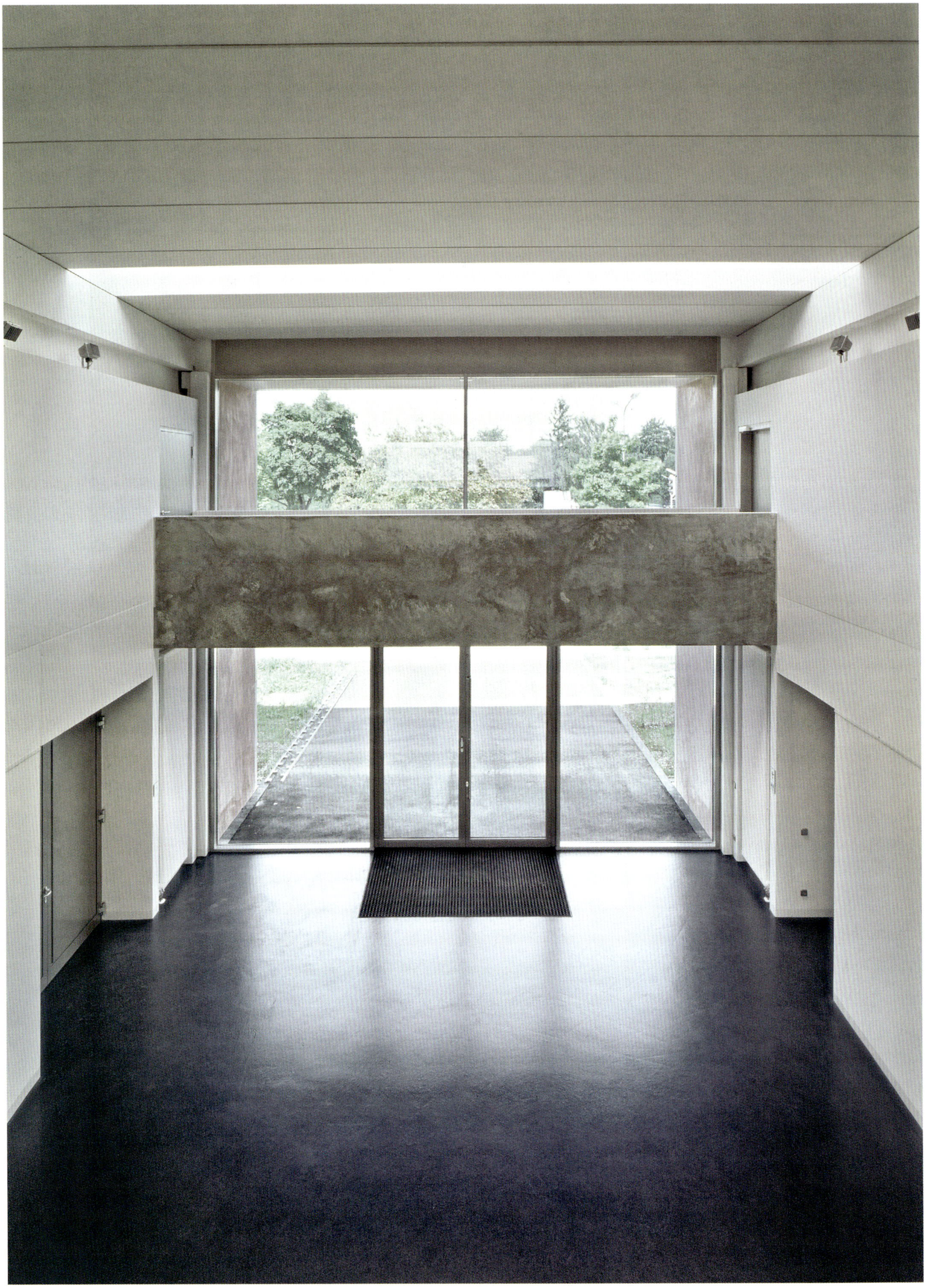

vertical section   scale 1:20

1 30 mm polycarbonate cellular skylight strip
2 80 mm planted layer with sedum
   20 mm drainage layer, separating layer
   160 mm insulation, vapour barrier
   8300 × 1200 × 200 mm hollow tensioned concrete
   50 mm insulated acoustic ceiling
   12.5 mm plasterboard
3 reinforced concrete beam, top graded to falls
4 wall floodlight in foyer
5 200 mm precast concrete wall element
6 sandwich element:
   80 mm reinforced concrete, pigmented
   100 mm insulation,
   140 mm reinforced concrete
7 floor construction:
   10 mm needle felt, 50 mm screed
   separation layer
   2× 22 mm impact-sound insulation
   220 mm reinforced concrete filigree slab
   services space, 12.5 mm plasterboard
8 aluminium sliding/bottom-hung casement
9 45/5 mm flat steel safety barrier
10 3 mm aluminium sheeting
   140 mm thermal insulation
11 foyer floor construction:
   30 mm asphalt, glass-fibre mat,
   160 mm reinforced concrete slab
   bituminous sheeting, 80 mm insulation
   50 mm foundation course, 200 mm gravel

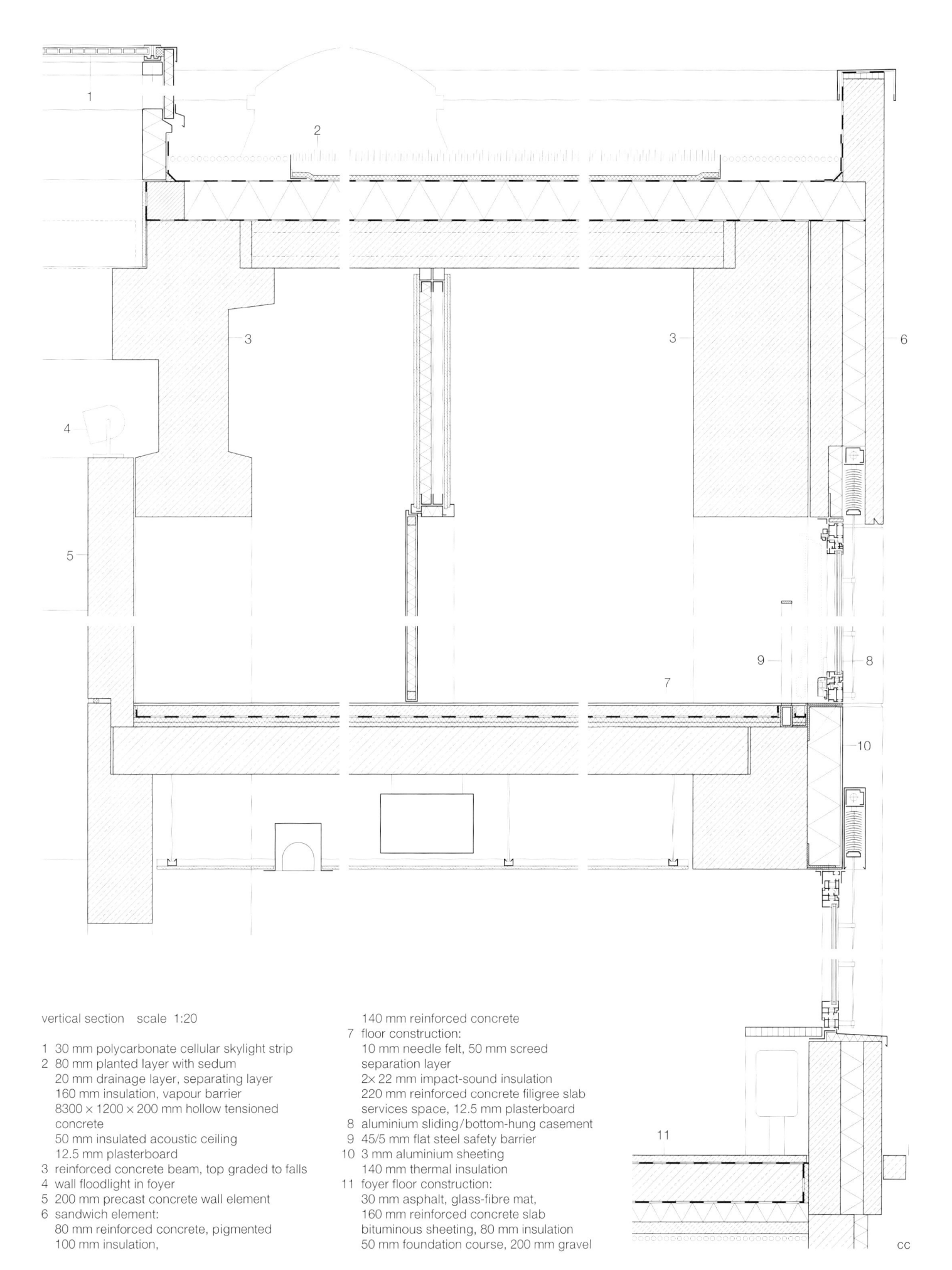

Cost-efficient building means sustainable building

Ingrid Amann

How, in your opinion, should the planning process be run in order to be as economical as possible?
Cost efficient building isn't just about economy. It is about economy in association with improved function and – most importantly – with high quality architectural design.
When these three criteria are met, then the fundamentals for sustainable and economical building have been achieved. With sustainability, I also include running costs and low energy consumption.
Even the topic of economy is not only referring to building as cheaply as possible, but rather, the creation of real estate, or city spaces which retain their value as long as possible. My focus is on architecture which offers usable, attractive spaces with good natural light and good artificial light; buildings which are appropriate for the location and for the users alike; architecture that one gladly calls to mind and that one cares and tends for. The use of resources, the whole effort involved; it has to be worth it from an energy and ecological point of view.

Bavaria's most cost efficient and most economical building – considered over a longer period of time – is Schloss Neuschwanstein, built by King Ludwig II. The fairytale castle, which is one of the most well known tourist attractions in Germany, is recognised world-wide as an icon of the German Romantic and attracts 1.3 million visitors annually. Once condemned by the nation and its citizens as shockingly wasteful, Neuschwanstein is now one of the most profitable constructions in the region. It is admittedly an exception, but it demonstrates the capriciousness of history.

How and in which phase of the planning process can you best reduce costs?
In the beginning it is all about finding out what the client wants and needs. As far as possible, the design must be appropriate to the function, the location and the limitations of the budget. Only then is it possible to determine what the essentials are, and to consciously experiment with them. We as architects are capable of adapting to the desires of our clients throughout the entire building process, and as such can – in association with the client – allocate funds in such a way as to optimise even the most modest of budgets. That means that funds should be spent where they most benefit the client and his construction, whether aesthetically or functionally.

Which areas offer the most potential for saving costs?
In my experience, prefabricated concrete construction can be much more economical than other comparable construction systems. It is essential to ensure the early collaboration with not only technical planners, but also the contracted building companies. When the contractors have the opportunity of working with the planners at an early stage – for example, on the pre-cast concrete facade panels for the advertising agency in Riem or the cultural centre in Trudering – then production requirements can be determined early, misunderstandings prevented, and costly and time consuming alterations largely avoided.

An optimal, continuous workflow during the entire planning and construction process is important. When technical engineers specify particular technical services or specific lighting, for example, the same engineers should be commissioned to oversee the site management of these components in order to secure the quality of the work.
An architect has much more influence over the optimisation of the cost planning when more management control is included in the contractual agreement. The point is surely not only to deliver a theoretical concept, but to produce a construction and to coordinate all the variations which occur during the complex planning process.

Only when we architects remain involved well into the construction phase of a building, is it possible to ensure positive communication. That means that we can ideally be involved in aesthetic discussions and decisions regarding variations, and thereby avoiding a variety of glaring mistakes.

What contributions can consultants, for example structural engineers, make in cost efficient building?
The close collaboration with structural planners is a deciding factor in ambitious, experimental planning of prefabricated constructions. In the printing works in Riem we developed a type of "click" system out of concrete piles and panels, for a hall with three galleries and skylights above. In coordination with the structural engineer it was possible to determine the optimal dimensions for the production of the concrete elements, the transport and the assembly.
Another aspect of our work is the motivation of the various contractors and planners. They are more likely to produce higher quality work, or make an extra effort at no extra cost, when they realise that something extraordinary is being produced and that new experience and know-how will be attained. In particular projects, where neither money nor time are available, then it is necessary to employ unconventional solutions, as in the printing works in Riem (ill. 1). There the possibilities for standardised production were fully exhausted and the facade design and building physics were developed based upon the characteristics of the economical material selected. When the time frame is long enough it is possible to absorb unexpected situations and to plan more flexibly, without losing any of the architectural quality or incurring higher costs.
Of course, even the client, as an involved party, is responsible for various decisions; to date none of my clients have actively contributed to planning or construction.

Essential, however, is a transparent, open communication between all involved parties, but predominantly with the client. He is the one who must make decisions promptly. The aim should be to prepare the decision-making process, in order for cost optimisation of to be possible when quotations are requested.

Do you prefer more elaborate solutions in particular situations? Which compromises are you prepared to make?
My main interest is the interaction of the architectural concept with the possibilities of implementation. If you have influence over the planning and its implementation and can affect the alterations that occur within the process in a positive way, then it is seldom necessary to make compromises.

Do you intentionally select some expensive solutions and then save in other situations?
In the cultural centre in Trudering, for example, we did intentionally make some decisions which were important to the concept and added value to particular elements; the large window in the hall, the ceiling heights and the large sliding elements in the seminar rooms.
The stairs are contrastingly simple, yet attractive: exposed concrete, straight out of the factory. The users can experience an explicit value system of architectural elements.

That doesn't mean that it looks cheap or that the surfaces are poorly executed. It was planned and specified from the very beginning that the pre-cast stairs should have smooth surfaces, sharp edges and slip-proof structure in the treads, as well as aesthetically detailed butt jointing – all that in only one stage of casting, without the need for screed, tiling or painting. It was necessary, however, to exclude any unexpected tolerances or prevent damage on site.

Where exactly are the advantages of building with pre-cast elements, when the purpose is to save costs?
Pre-cast elements have the advantage of prefabrication over in-situ concrete. The elements are manufactured under perfect conditions in the factory; formworking, mixing and surface treatment. The controlled climate also ensures superior surfaces to that of in-situ concrete.

How do you see the future of cost and resource efficient building?
The future of building will be determined by ecological and economical factors during planning, construction and operation of the building. The architect must re-evaluate his position and search for new possibilities in order to be knowledgeable in this complex matter.

Construction know-how, materials and process alternatives are enormously important. In addition to which comes the knowledge of the half-finished products and standard components which are available on the market today, and their application. The willingness to experiment is promoted by the true understanding of architectural matter. It will become even more important, in both research and education, to analyse and develop the ecological and economical conditions of construction.

What importance do you place on the future running costs of a building when planning?
Clear analysis of the way current savings may incur future running costs, or vice versa, is an essential element of communication with the client. It is our responsibility to determine and research what is the most sensible solution and which architectural materials, for example skylights for daylight or high quality surfaces which do not require painting or elaborate cleaning, are capable of keeping future running costs to a minimum.

1 Printing works in Munich-Riem, 1998

Secondary School in Brixlegg

Architect: Raimund Rainer, Innsbruck

"Passive house" standards with low running costs
Cost saving through low energy consumption
Atrium as central meeting place

The construction of this new secondary school in Brixlegg was not directed by cost savings; rather it was based upon the desire of the Tyrolean municipality to act in an environmentally sustainable manner at the communal level. While constructed at little more than the cost of a conventional school building, the running costs are considerably less due to selected construction techniques; the annual oil consumption is approximately a third of that of a conventional school building.

Arrangement
Access to the area for all-day tuition on the ground floor is via the set-back, covered main entrance from the playground. From here, a staircase leads up to the heart of the building: a large, top-lit atrium around which the classrooms are laid out on three levels. This "square" is encircled by the solid buildings which accommodate classrooms and group study rooms. "Lanes" between the structures create links with the outside world and provide cloakrooms for the pupils.

Methods
The school is constructed of reinforced concrete with a composite thermal insulation system applied externally. With an air-tight built envelope and a central ventilation plant with heat recovery system, the building complies with recommended passive-energy standards. In order to fulfil these requirements, technical specialists were involved in the planning process from the preliminary stage onward.
In winter, pre-heated air is blown into the classrooms via inlets in the lower part of the facade. Exhaust air is extracted beneath the soffit through sound absorbers into the corridors on each level and expelled out via a central shaft. Thermal energy from the double cross-flow heat-exchange unit is used to warm the incoming air. Solar collectors, with a total area of 200 m², provide most of the heating energy required between seasons, in addition to supplying heat to the local swimming pool. An existing oil-powered heating plant in the neighbouring polytechnic meets residual heating needs via simple radiators. Automatically controlled external louvers shade the classrooms and internal sun sails beneath the sky lights protect the atrium. In summer, the in-coming fresh air is evaporatively-cooled. The ventilation plant also operates outside school hours and provides additional night-time cooling. Finally, the thermal storage capacity of the solid concrete structure helps to ensure an even indoor climate all year round.

Project details:
Usage: school with all-day tuition
Construction: reinforced concrete
Access: atrium
Internal ceiling height: 3.2–3.7 m
Built site area: 925 m²
Total built area: 4,100 m²
Total internal volume: 16,245 m³
Usable area: 3,441 m²
Construction cost: 4 million Euros
Cost/m² usable area: 1,162 Euros
Site area: 6,229 m²
Heating requirements: < 15 kWh/m²a
Period of construction: 11 months
Date of completion: February 2007

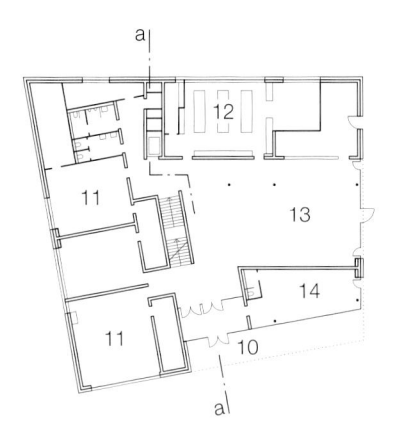

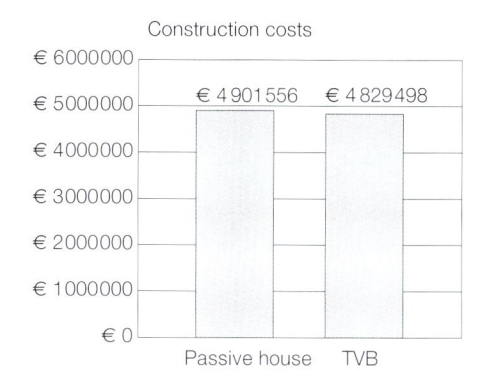

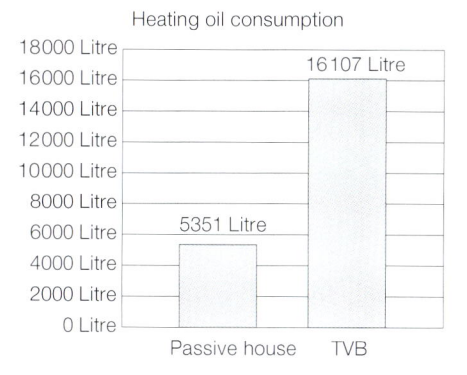

Table 1: Comparison of construction costs and heating oil consumption of a passive house with those of a building constructed to local standards (TVB).

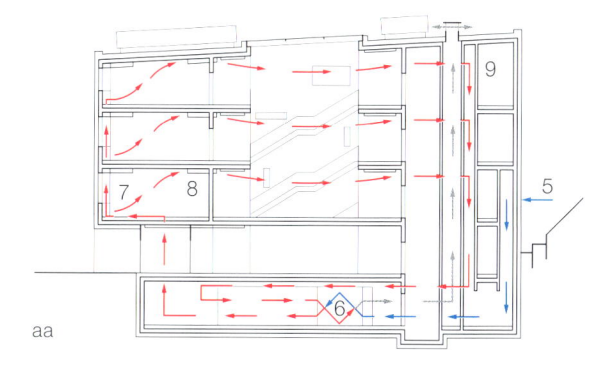

site plan
scale 1:2500
floor plans:
ground, first, second
scale 1:750
section ventilation diagram (winter)
scale 1:500

1 secondary school
2 primary school
3 polytechnic
4 kindergarten
5 fresh air intake: first floor facade
6 ventilation plant with double cross-flow heat-exchange unit
7 fresh air flow in plinth zone
8 high-level air flow in classroom/corridor with sound absorber
9 central extraction at soffit zone
10 main entrance
11 specialist teaching
12 school kitchen
13 full-day school area/assembly hall
14 recreation space for pupils
15 conference room
16 principal
17 assembly hall
18 group room
19 pupils' cloakroom
20 classroom
21 library

1
2

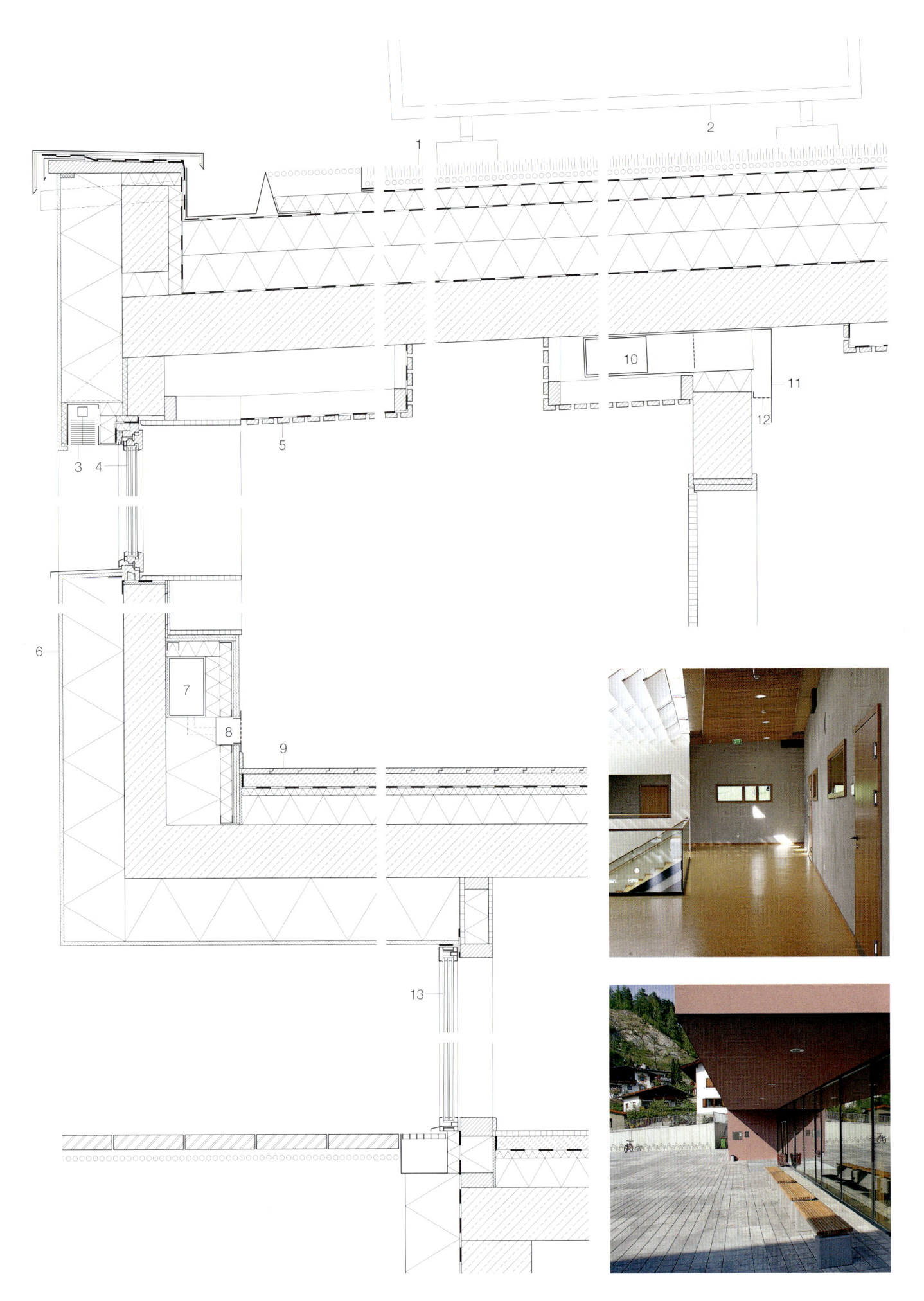

3 4
5

10
11
12

6
7
8
9

13

116

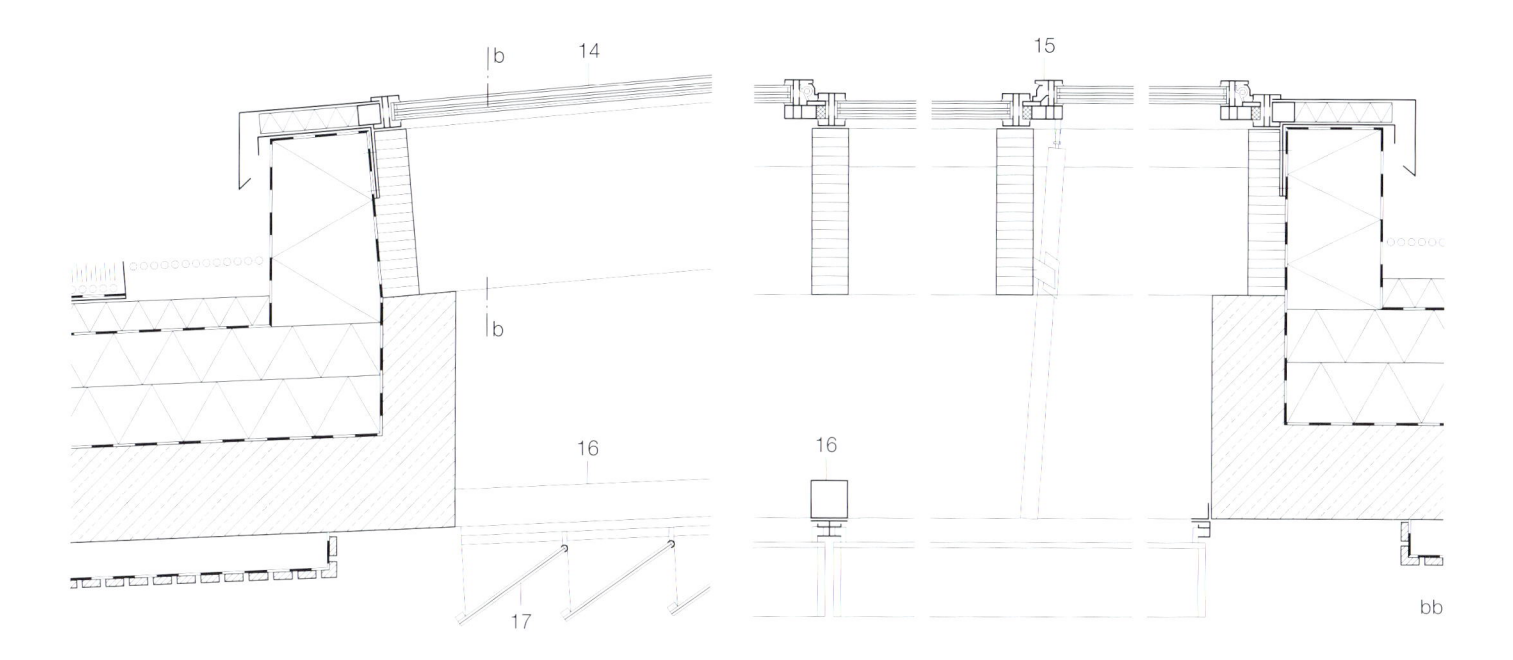

vertical section
north-west facade with skylight
scale 1:20

1 roof construction (U = 0.10 W/m²K):
 100 mm extensive roof planting
 drainage and storage mat
 60 mm rigid-foam insulation
 three-layer bitum. seal with
 copper inlay
 300 mm two-layer EPS thermal
 insulation
 bituminous vapour barrier,
 undercoat
 250 mm reinforced concrete
 roof slab
2 solar collector
3 louver sunblind
4 casement (UW = 0.79 W/m²K):
 triple glazing in thermally separated
 aluminium/timber frame
5 suspended acoustic ceiling
 solid oak strips, acoustic mat
6 wall construction
 (U = 0.13 W/m²K):
 painted lotus effect on silicone-resin render
 260 mm EPS thermal insulation
 180 mm reinforced concrete
 290 mm cavity insulation
 25 mm two-layer plasterboard
7 160 × 250 mm air-supply duct with
 integrated sound absorbers
8 fresh-air inlet
9 22 mm oak-strip parquet flooring
10 high-level air duct to classroom/corridor
 with integrated sound-absorbers
11 sheet-steel lining
12 perforated sheet-metal air outlet
13 triple glazing (UG = 0.6 W/m²K)
14 roof light with triple glazing
15 smoke and heat exhaust flap
16 100 × 100 mm steel SHS downstand beam
17 adjustable fabric sunblind

Secondary School in Eching

Architects: Diezinger & Kramer, Eichstätt

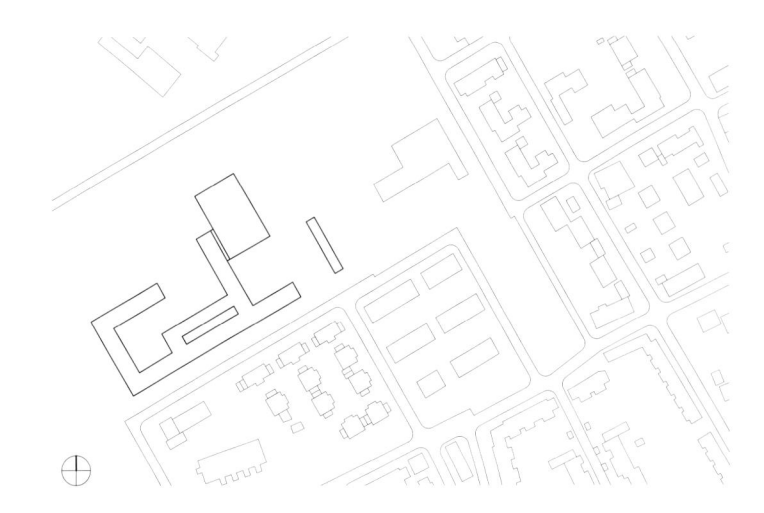

Limited range of construction elements
Intentional selection of medium quality materials

Planned for roughly 1,000 pupils, this new secondary school is situated on the periphery of Eching, north of Munich. Because of its somewhat isolated location between a suburban railway station and a residential area, some distance from the town centre, the architects decided to design the school as an autonomous urban entity with a strong relationship to the external surroundings.

Positive atmosphere created by colour concept

The layout comprises a series of tracts placed at right angles to each other, a result of which is that the ensemble appears to consist of a number of separate buildings. The tracts encompass two courtyards – a forecourt and a playground – each of which is defined on the open side by a series of steps. These two external spaces are linked by a central hall, which forms the heart of the complex and is used as a recreation space during school breaks, a venue for various events, a foyer, meeting place and circulation space. The intersection of the hall with the two upper floors – in combination with a large skylight – creates a generous, open environment. All areas of the school are easily accessible from the central hall: the administration, sports hall, classrooms and special teaching facilities. The adjoining sports hall is recessed one level into the ground in order to be compatible with the scale of the surroundings. The classrooms are oriented to the south, while the special teaching facilities face west and north in order to benefit from diffuse natural lighting. In these areas of the school the corridors are enlivened with spatial projections and recesses, creating intimate spaces that can be used during breaks as informal meeting places for the pupils.

The appearance of the school building is distinguished primarily by the striking colours of the rendered facades – lilac tones for the classrooms and yellowish green in the corridors. This coloration is continued internally in the floor finishes. An additional positive effect of this colour concept is that the otherwise austere, white surfaces are suffused in a soft light.

Cost efficiency

It was possible to minimise costs by using simple forms of construction and reasonably priced materials, and by limiting the range of window and door formats. Precisely detailed specifications also allowed expensive alterations to be avoided during the construction phase. The architects deliberately selected more expensive materials or techniques in particular situations in order to heighten the perceived quality of the building and to guarantee the desired longevity; for example, the internal sprayed render and the metal work on the balustrades.

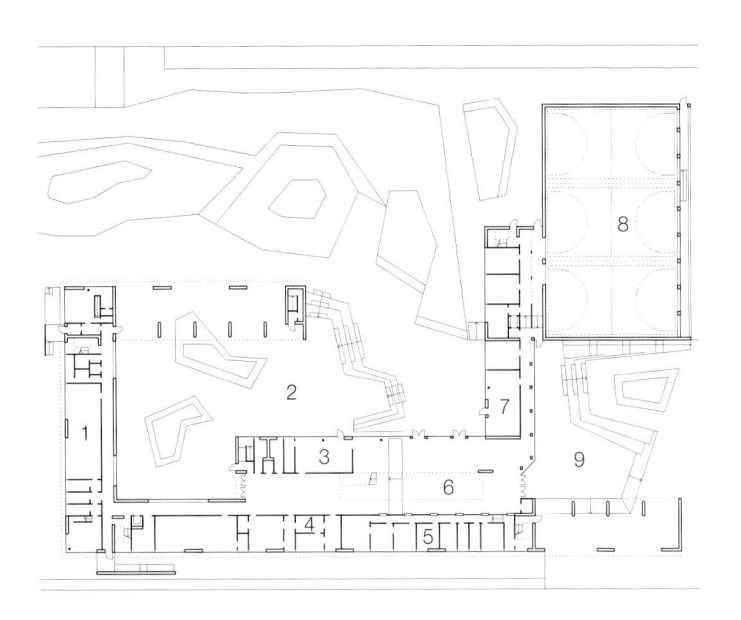

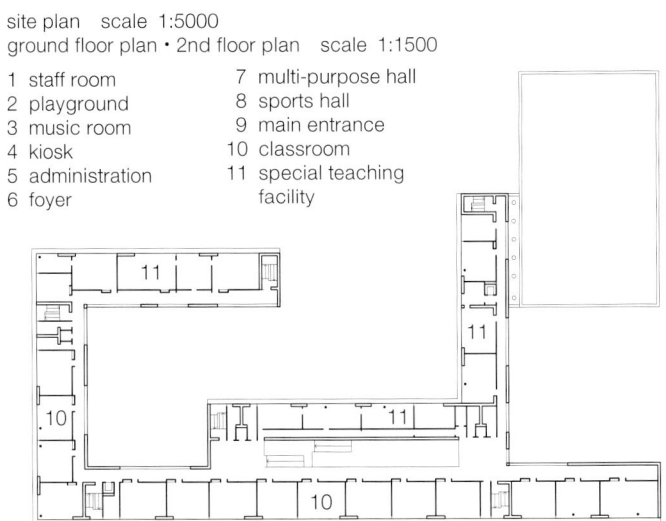

site plan   scale 1:5000
ground floor plan • 2nd floor plan   scale 1:1500

1 staff room
2 playground
3 music room
4 kiosk
5 administration
6 foyer
7 multi-purpose hall
8 sports hall
9 main entrance
10 classroom
11 special teaching facility

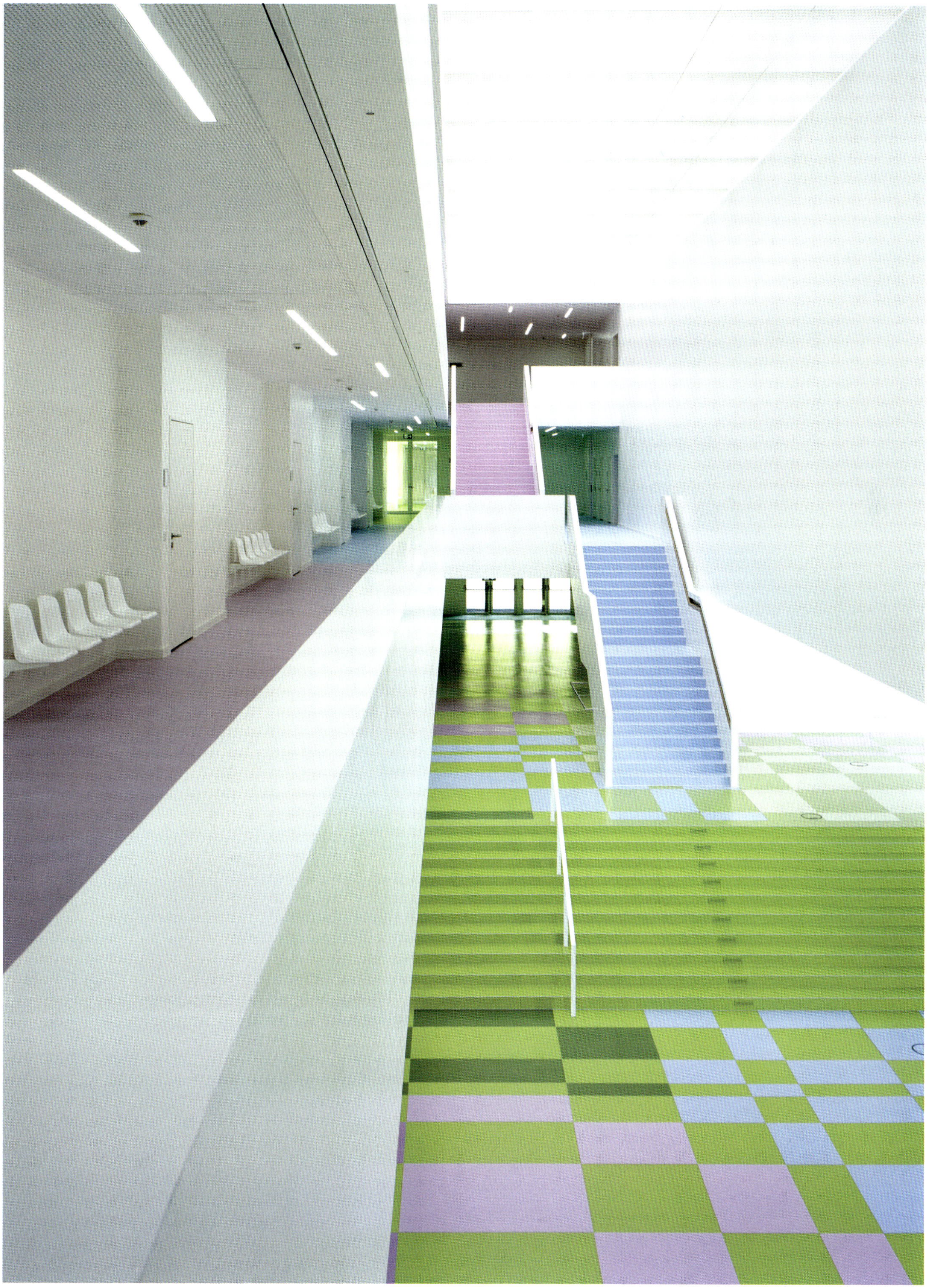

vertical section corridor facade
vertical section classroom facade
scale 1:20

3 timber and aluminium window
   with double glazing
1 window reveal with latex coating
2 110 mm composite thermal insulation
   system
   250 mm reinforced concrete wall
4 aluminium section window sill
5 6 mm sheet-steel window sill, painted
6 100 × 50 × 3 mm steel RHS
7 2 mm galvanized sheet steel
8 recess for light fitting
9 foot wiping zone: doormat
13 4 mm sheet aluminium fascia
12 pre-cast concrete element with
   thermally separated reinforcement
   connection

11 void closed at side
10 roof construction: 50 mm gravel
   0.4 mm sheet stainless steel sealing
   layer, line welded
   120 mm expanded polystyrene thermal
   insulation, vapour barrier
   4 mm bitumen sealing layer
   bituminous undercoat
   260 mm reinforced concrete roof to falls
   suspended acoustic ceiling
14 sliding window with double glazing
15 2 mm sheet aluminium panel
16 floor construction: 2.5 mm floor finish
   60 mm cement screed
   PE membrane
   40 mm impact-sound insulation
   PE membrane
   260 mm reinforced concrete floor
   suspended acoustic ceiling
17 fixing for sunshade cable, drilled

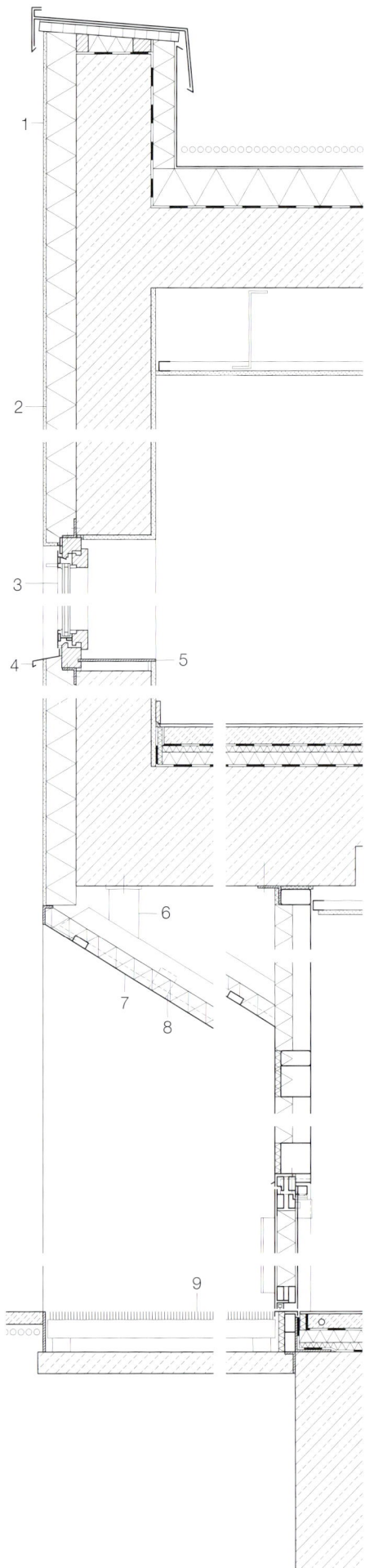

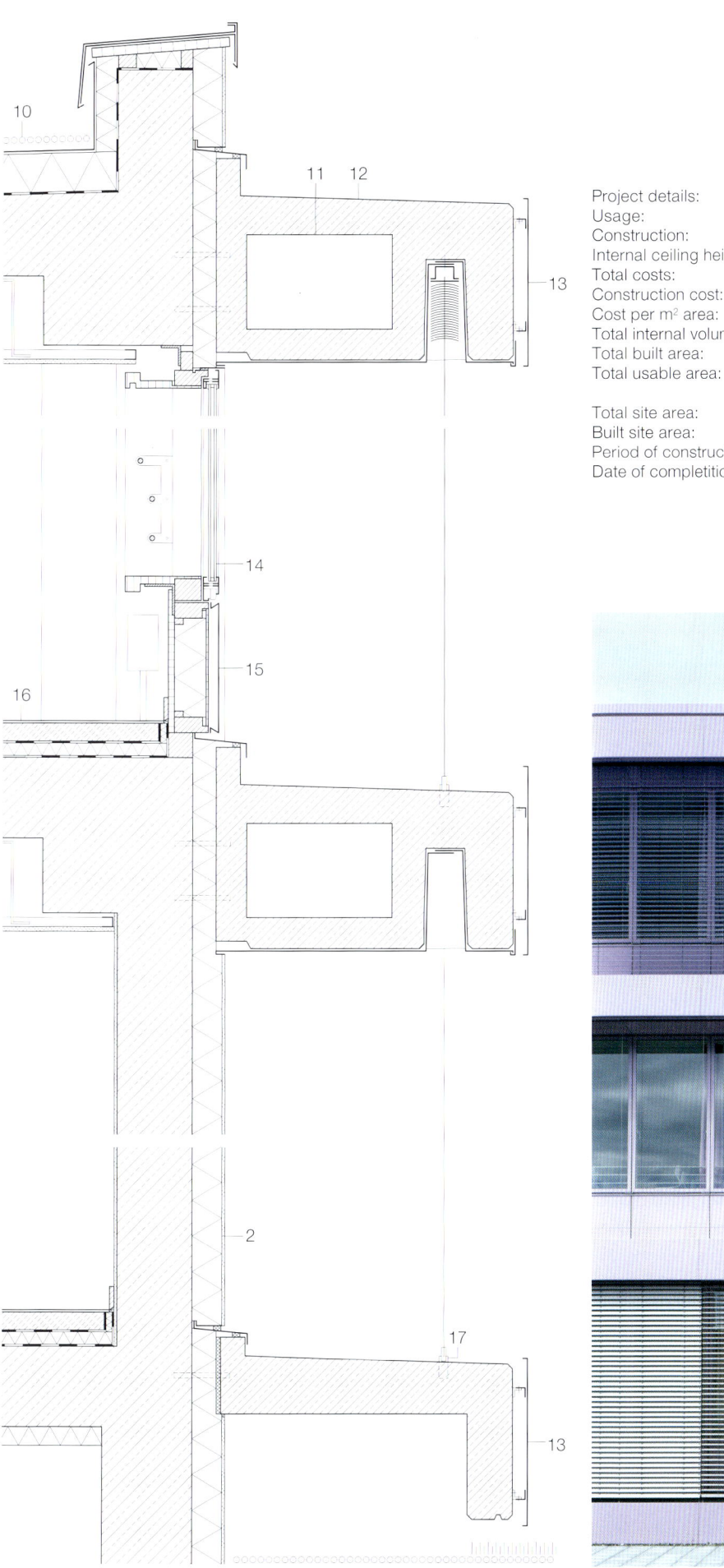

Project details:
Usage: secondary school
Construction: reinforced concrete
Internal ceiling height, classrooms: 3.1 m
Total costs: 21.8 mill. Euros
Construction cost: 13.8 million Euros
Cost per m² area: 1,090 Euros
Total internal volume: 59,580 m³
Total built area: 12,630 m²
Total usable area: 5,464 m² (school)
1,738 m² (sports hall)
Total site area: 20,263 m²
Built site area: 5,503 m²
Period of construction: 18 months
Date of completition: Aug 2006

Montessori College Oost in Amsterdam

Architects: Architectuurstudio Herman Hertzberger, Amsterdam

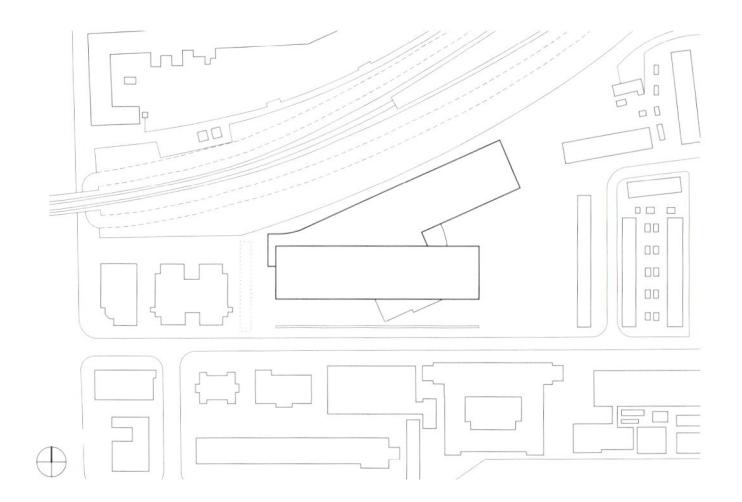

Microcity with high recreational quality
Circulation and visual interaction

In the Montessori College Oost, the great hall interconnects all the various elements of the school and takes on the function of a central square in a town; a meeting place for pupils and teachers before, between and after classes. The hall connects the lower wing, which accommodates the specialised teaching facilities, with the larger five-storey tract where the classrooms are located. The hall subdivides the taller tract into two sections which are offset from each other in height in a split-level manner by half a storey. The linking staircases to the upper levels are suspended in the air like bridges and placed in such a way that they are not directly above each other, but rather enable visual links to be maintained. These "staircase bridges" are not only access routes but also provide relaxation and communication zones. There are steps which the pupils can use to sit down or even write on, and fold-down tables have been incorporated into the balcony balustrades.

Ingenious planning was required to create an uninterrupted hall space in a building of this size without having to divide it into various fire compartments by creating alternative escape routes. From the classrooms, access is provided to the enclosed fire-escape stairs via external balconies along the facades. The unobtrusive staircase structures also act as transverse bracing in the two sections of the tall tract, while the balconies provide sun shading for the facades. The slender, elongated form of the building is emphasised by the load-bearing structure, which is longitudinal rather than transverse.

Acoustics and materials

The concept of the great hall forced the architects to address the question of acoustics in order to achieve undisturbed classrooms on the one hand while still providing a space capable of hosting large events on the other. The solution to the problem was found in the detailing. The walls facing into the atrium are clad with felt, and the stairs in this space are lined with timber acoustic panels. The architect simultaneously sought to achieve a balance between "hard" and "soft" materials. Sheet-steel cladding was used for balustrade panels, in addition to other areas, in order to contrast with the felt and timber. The same philosophy led the architect to decide on the "facade roof" which, to a certain extent, creates the framework for the glass facade.

Project details:
Usage: Montessori school
Construction: reinforced concrete
Internal ceiling height: 2.1 m classrooms
2.25 m corridors

Number of storeys: 4 split levels
Total cost of construction: 15,158,200 Euros
Cost per m² usable area: 891 Euros
Room sizes: 100 m² workshops
64 m² specialised teaching
40 m² classrooms
Total usable area: 17,016 m²
Sports area: 315 m²
Period of construction: 19 months
Date of completion: October 1999

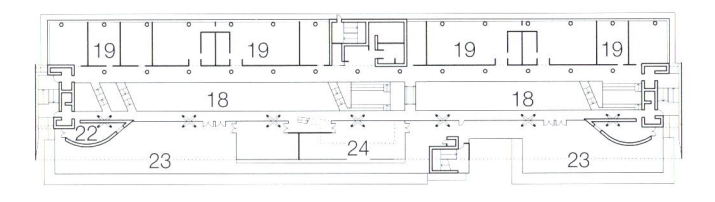

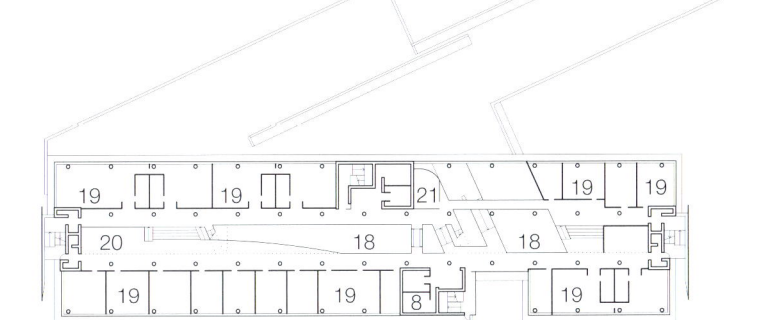

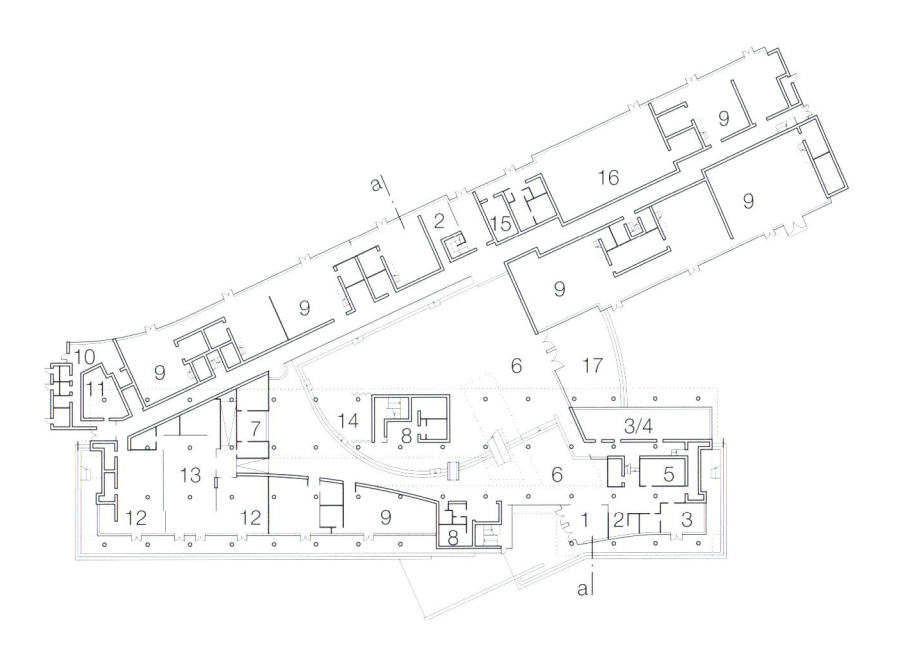

site plan
scale 1:4000

fourth floor plan
first floor plan
ground floor plan
scale 1:1250

1 entrance
2 reception
3 staff
4 office
5 store
6 hall
7 buffet
8 WC
9 workshop
10 printing shop
11 services
12 training kitchen
13 training restaurant
14 film screenings
15 change room
16 sports hall
17 recreation area
18 void
19 classroom
20 gallery
21 staff room
22 air conditioning
23 roof terrace
24 library

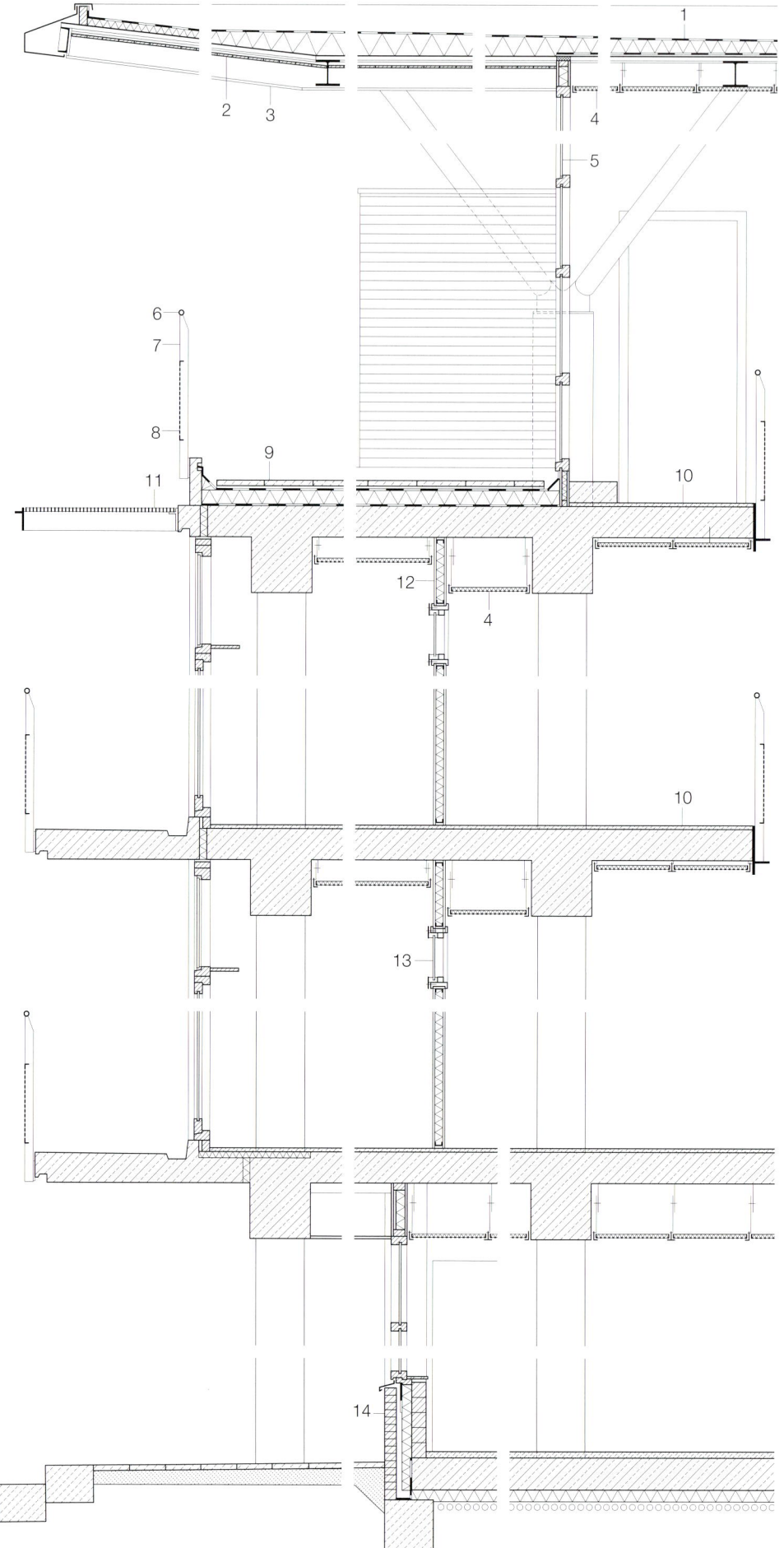

vertical section
scale 1:50

1 roof construction:
   bituminous sheet sealing layer
   180–120 mm mineral-wool thermal insulation, finished to falls
   vapour barrier
   35 mm ribbed metal sheeting
   200 mm steel I-beam
2 18 mm red cedar soffit lining
3 240 mm steel I-beam
4 30 mm mineral-fibre sheeting
5 double glazing in iroko frame
6 Ø 34 mm tubular steel handrail
7 20 × 70 mm steel flat post
8 3 mm perforated sheet-steel panel
9 roof terrace floor construction:
   600 × 600 mm concrete paving slabs on neoprene bearers
   bituminous sheet sealing layer
   120 mm rigid-foam thermal insulation
   vapour barrier
   250 mm reinforced concrete slab
10 floor construction:
   3 mm PVC flooring
   30 mm screed
   250 mm reinforced concrete slab
11 30 mm galvanized steel grating
12 70 mm channel section stud-wall system with
   70 mm mineral-wool thermal insulation and
   12.5 mm plasterboard lining on both faces
13 7 mm wired glass in frame, consisting of
   5 × 30 mm steel flats and
   50 × 18 mm MDF strips
14 wall construction:
   95 mm black brick facing skin
   45 mm ventilation cavity
   80 mm mineral-wool thermal insulation
   150 mm sand-lime brickwork

section
scale 1:600

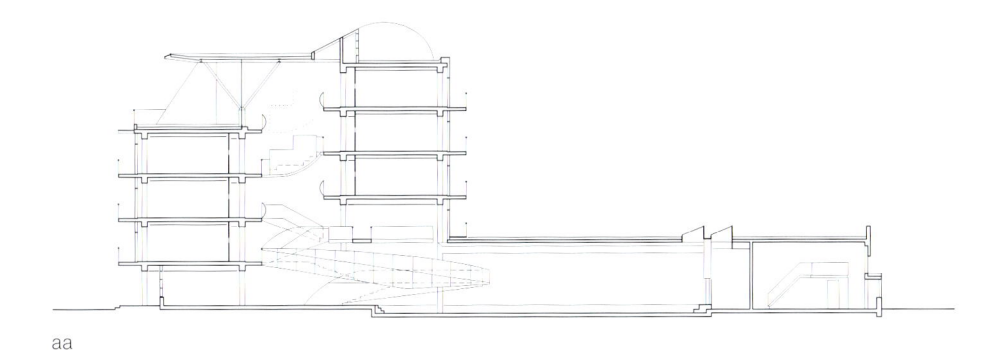

aa

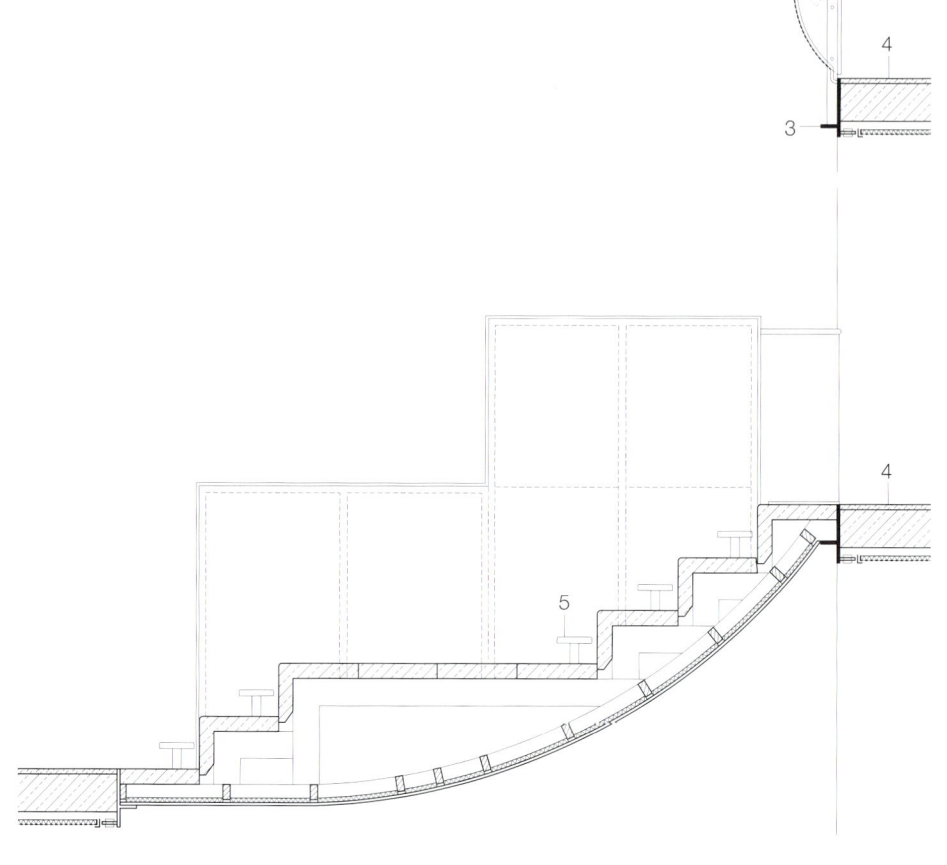

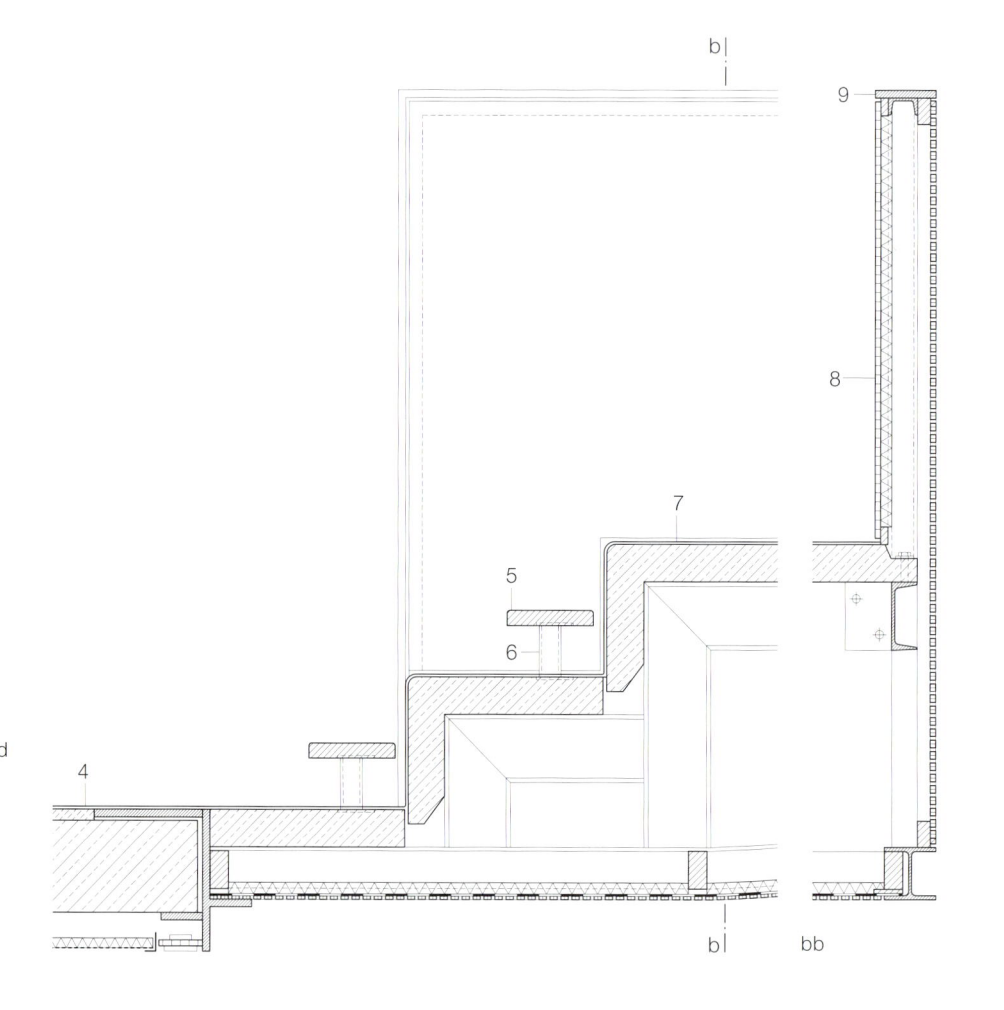

vertical section
scale 1:50
detail section
scale 1:20

1 3 mm perforated sheet steel
2 folding table:
   40 mm laminated construction board
3 120 × 10 mm and
   400 × 10 mm welded
   steel flats
4 floor construction:
   3 mm PVC flooring on 30 mm screed
   250 mm reinforced concrete slab
   30 mm mineral-fibre panels
5 245 × 38 mm iroko treads
6 Ø 65 mm steel tube
7 2.5 mm linoleum on 100 mm pre-cast
   concrete panel
   180 mm steel channel sections
   30 mm rockwool sound insulation
   50 × 100 mm softwood battens
   140 mm steel I-sections, fire-resisting mat
   9 mm perforated laminated construction board
8 15 mm laminated construction board
   30 mm rockwool sound insulation
   20 × 40 mm okumén timber battens
   80 mm steel channel sections
   30 × 60 mm okumén timber battens
   15 mm perforated laminated
   construction board
9 165 × 20 mm steel plate

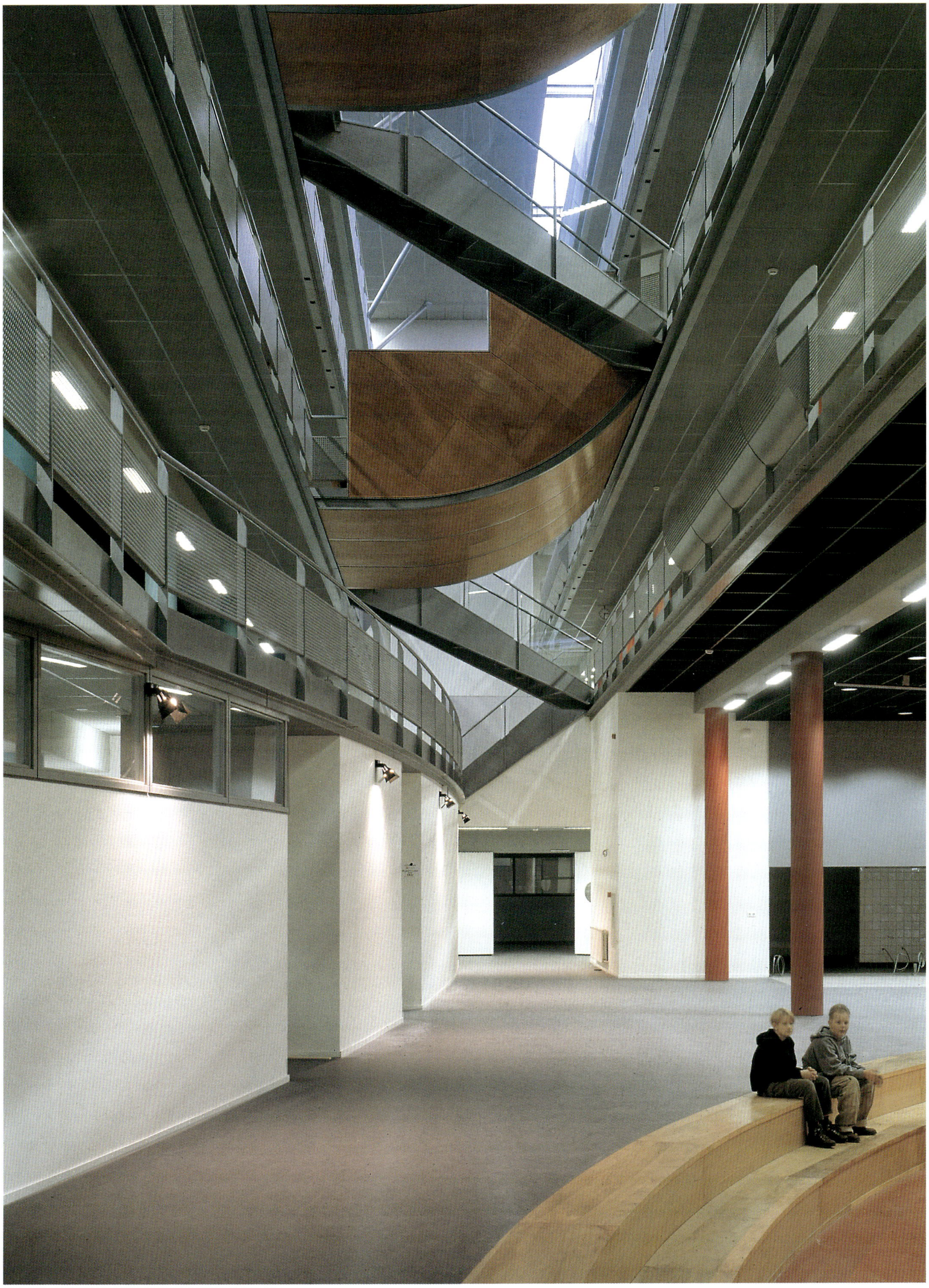

Ice Skating and Mini-golf Hall in Bergheim

Architects: mfgarchitekten, Graz

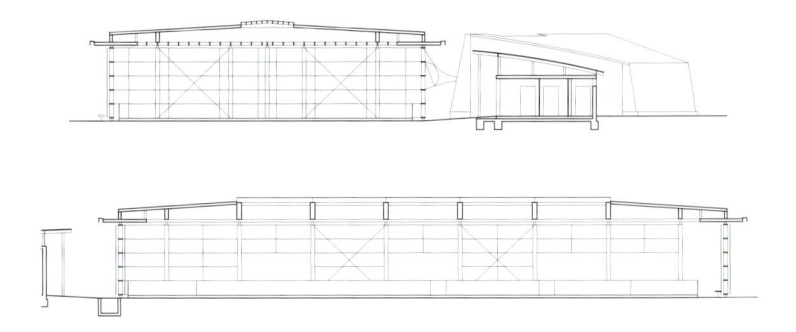

Spanning a 20 x 40 metre space of varying functions

Between the existing leisure activity constructions in Bergheim, near Salzburg, the clean-cut structure of this new hall provides a central calming influence. The 20 × 40 metre space alternates function every six months; in summer it offers mini-golfers shelter from sun, wind and weather, while in winter the ice skaters benefit from its protection.

Externally mounted timber structure with fabric sunscreen

An external polyester fabric covering and homogenous timber soffit grid, located beneath the large roof light, serve to diffuse the incoming daylight and to eliminate glare. The artificial lighting is made up of both indirect lighting through the roof panels and direct lighting on the function spaces. At night the construction transforms itself into a glowing object with opaque, shimmering skin. The rhythmical, articulated facade can be partially slid open, allowing the internal space to merge with the landscape.

The structural system of the roof, which is concealed beneath the timber grid, consists of laminated timber beams locally braced at the perimeter by laminated timber panels. This system is supported on a grid of hinged columns with individual cross-braced bays. The timber soffit grid was left in a rough sawn state and has a high moisture-absorbing capacity which helps to avoid secondary condensation in winter, as the elements cool. The timber structure remains dry and free of mildew due to the warmth of the sun and artificial lighting, in combination with natural ventilation.

Project details:
Usage: Ice skating and mini-golf hall
Construction: timber skeleton frame
Internal ceiling height: 5.0 m
Total construction cost: 368,000 Euros
Cost per m² usable area: 430 Euros
Total internal volume: 5,130 m³
Covered area: 886 m²
Usable area: 855 m²
Period of construction: 3 months
Date of construction: 2005

sections
scale 1:500
vertical sections · horizontal section
scale 1:20

1 27 mm larch plywood cladding
2 Ø 140–110 mm vacuum system rainwater collection pipe, heated
3 PVC roof sealant, mechanically fixed plastic fibre mat
   108 mm laminated softwood slab
4 2, 160 × 480–1,080 mm graded laminated timber beams
5 80 × 140 softwood purlin
   180–210 mm counter-beams
6 180 × 760 mm extruded corrugated polycarbonate slab
7 prefabricated soffit grid element
   50 × 30 mm sawn softwood battens, untreated
   20 × 80 mm counterbattens
8 50 × 160 mm softwood purlin
9 polyester sunscreen fabric on
   80 × 60 mm steel RHS frame
   2, 240 × 60 mm laminated louvers
10 Ø 10 mm galvanised steel suspension rod
11 80 × 60 mm galvanized steel RHS
12 access protection grating:
   double-bar galvanised steel mat
13 80 × 80 mm galvanised steel SHS column foot with 240 × 240 × 30 mm footplate
14 40 mm existing asphalt drainage layer
15 Ø 24 mm steel tension rod
16 240 × 240 mm laminated larch column

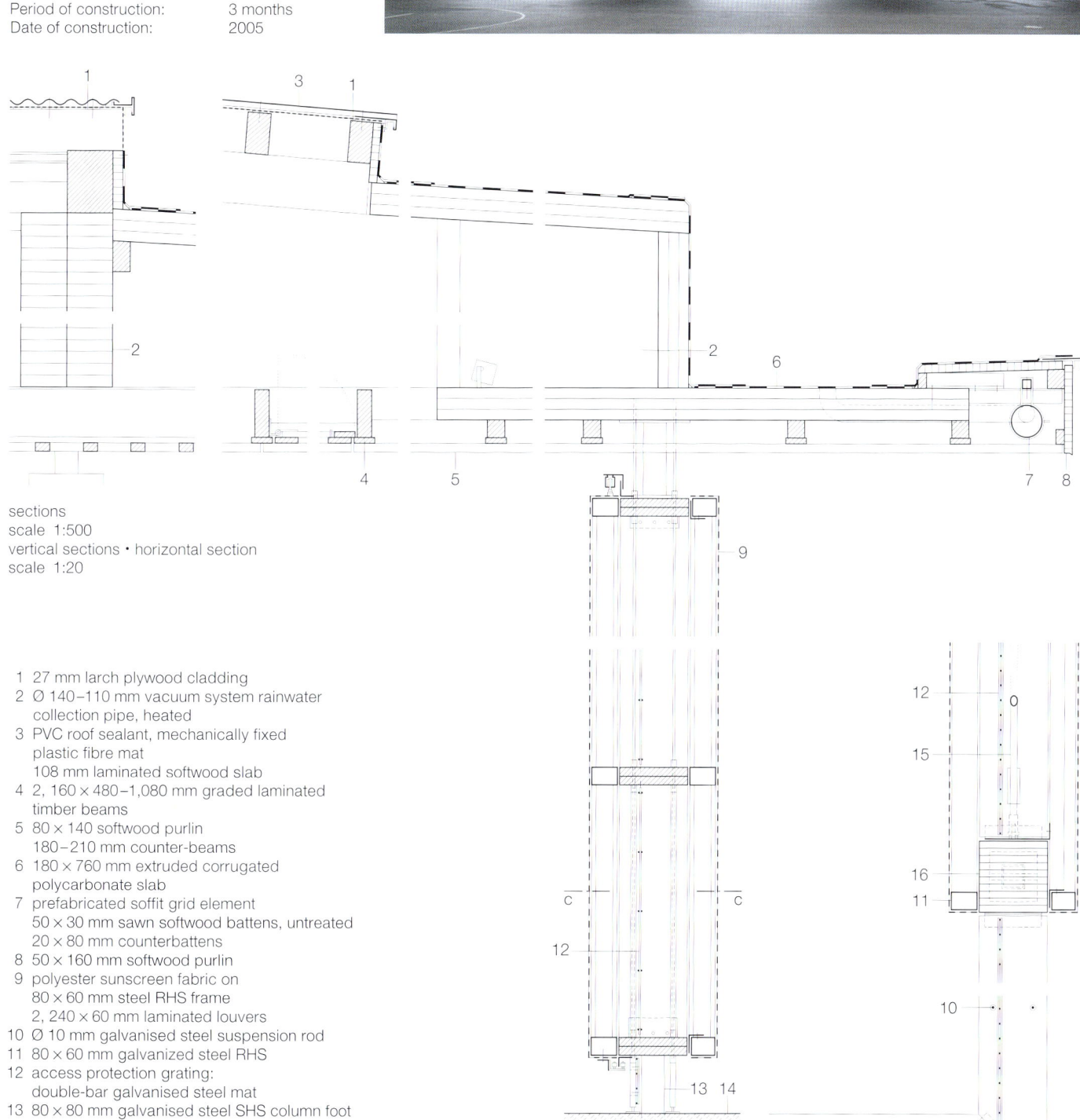

Ice Stadium in Wolfsburg

Architects: Schulitz + Partner, Braunschweig

Simultaneous design and detailing
Completion in nine months
Universally applied cost-saving measures

After gaining promotion to the top division in Germany, the Wolfsburg ice-hockey club was obliged to modernise its stadium in order to comply with the requirements of the sport's governing body, the German Ice Hockey League (DEL). The existing arena, dating from the 1980s, no longer met these standards. The first proposal, which was backed by an investor for a cost of 26 million Euros, unfortunately collapsed causing the club to miss the first season. In order to remedy this as quickly as possible, a new ice stadium had to be erected within ten months. In order to maintain the project, the city of Wolfsburg leaped into the financial breach with a budget of 7.5 million Euros.

Saving potential in planning and specifications

The architects carried out the design and construction planning simultaneously, so that it was possible to go out to tender with virtually no delay. Cost savings were achieved through a functional and constructional minimization of the planning that already existed. Incorporating a very small part of the existing hall, the original design concept was simplified, and the foyer and VIP areas were pared down. The curved walls, originally planned as in-situ concrete were redesigned as partially prefabricated polygonal elements. Screed has been done away with wherever possible, with due consideration to sound protection, and the concrete floor slabs simply trowelled and coated. The roof structure, with an optimized moment curve, was designed as a timber or steel construction. Both designs were put out to tender, the more economical solution being finally selected, namely that in steel. All possibilities for cutting costs were exploited.

Articulated structure

In order to provide the hall with a more dynamic appearance, the architects articulated the large-scale, closed external surfaces. Instead of selecting a uniform, rear-ventilated, sheet-metal facade, a structure was created at no additional cost which consists of aluminium elements of equal depth but varying widths. These panels enliven the facade of the building and, in certain light, lend it a undulating, wave-like appearance. Additional requirements for fire protection and mechanical services could be offset by making savings elsewhere. Nine months after the commencement of planning and six months after construction began; the ice stadium was handed over in time for the start of the season.

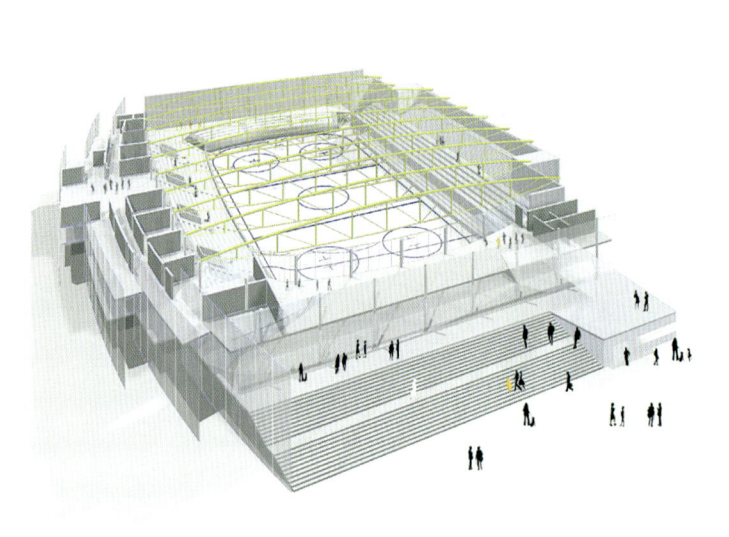

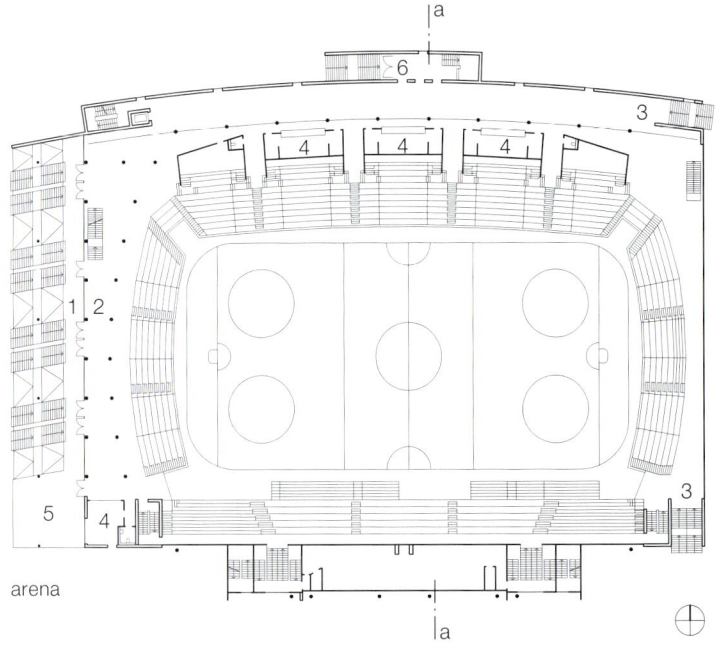

aa A

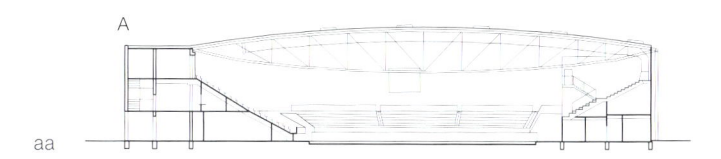

floor plans • section
scale 1:1000

1 main entrance
2 foyer
3 side entrance
4 kiosk
5 balcony
6 VIP entrance
7 VIP lounge
8 kitchen
9 seating

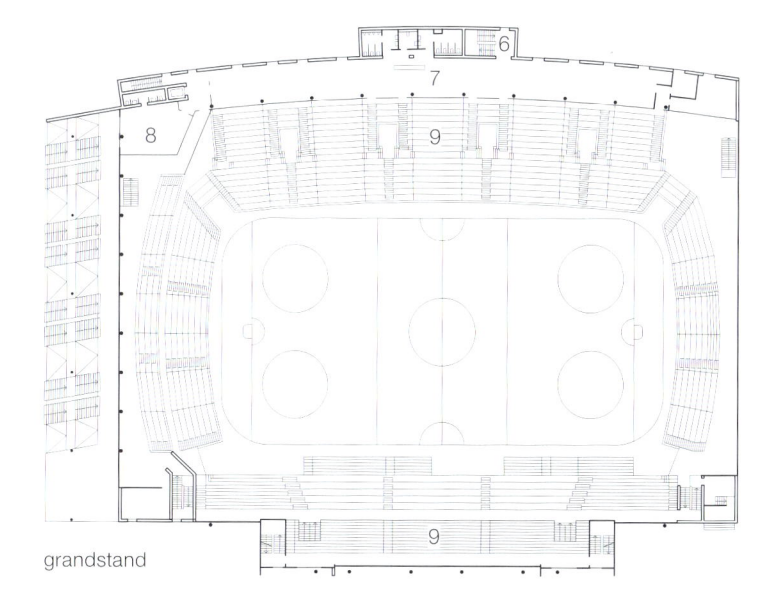

grandstand

Project details:	Ice stadium
Construction:	steel
Construction cost:	8.7 million Euros
Cost per m² area:	825 Euros
Total floor area:	10,540 m²
Total internal volume:	76,720 m³
Date of construction:	2006

1 2 mm sheet-aluminium cladding
2 3 mm sealing layer
   200–150 mm expanded polystyrene insulation
   2 mm vapour barrier, 250 mm r. c. slab
3 double glazing in alumin. frame (U = 1.8 W/m²K)
4 2 mm sheet-aluminium cladding
5 8 mm carpet, 46 mm screed
   separating layer
   250 mm reinforced concrete floor slab
6 3 mm sealing layer, 120 mm thermal insulation
   vapour barrier
   106 mm trapezoidal-section metal sheeting
7 240 mm steel I-section roof beam
8 300 mm steel I-section upper chord
9 220 mm steel I-section lower chord
10 steel connecting plate bolted to
   220 mm steel I-column
11 steel connecting flange
12 180 × 240 × 20 mm steel bearing plate on
   220 × 360 × 40 mm poured concrete pad
13 steel seating on Ø 406/12.5 mm steel column
14 2 mm sheet-aluminium facade section
15 65 mm ribbed sheet-aluminium section
   40 mm ventilated cavity
   100 mm sheet-steel fixing brackets
   100 mm mineral-wool thermal insulation
   200 mm reinforced concrete wall
16 coating on power-trowelled concrete
17 200 mm steel I-section column
18 sandwich panel: 2 mm sheet aluminium
   60 mm mineral-wool thermal insulation
   2 mm sheet aluminium
19 steel angle facade closing strip
20 10 mm rubber flooring, 46 mm screed
   separating layer, 150 mm reinforced
   concrete slab, 50 mm lean concrete,
   separating layer, 100 mm thermal insulation
   300 mm gravel

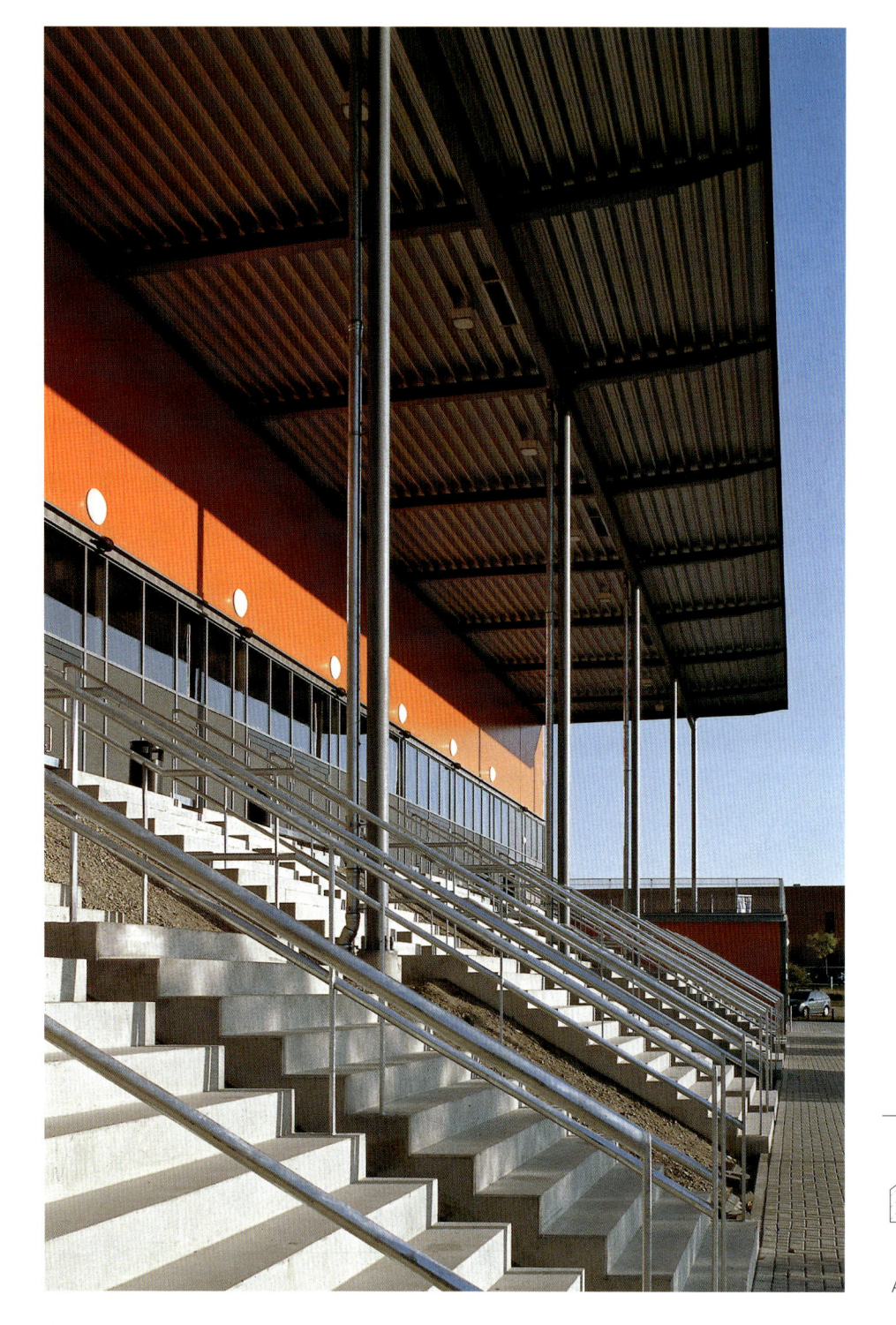

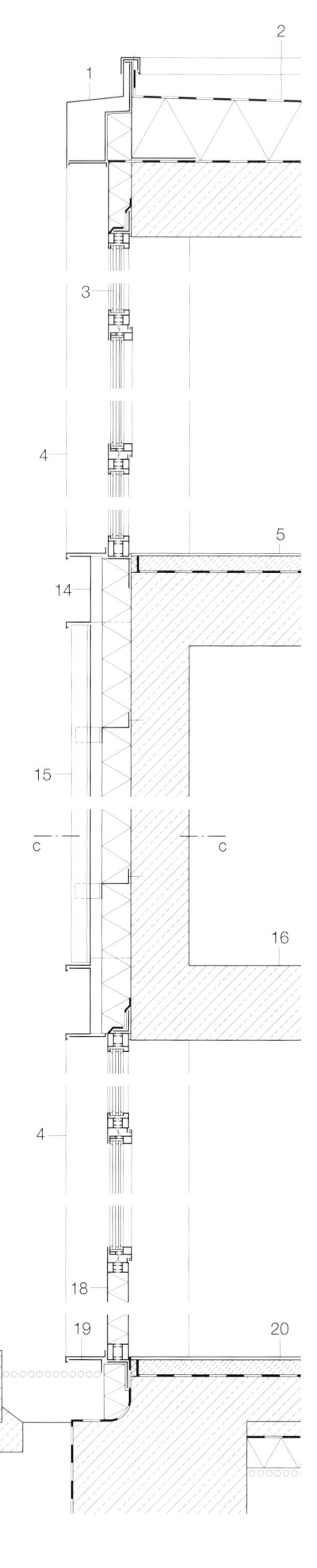

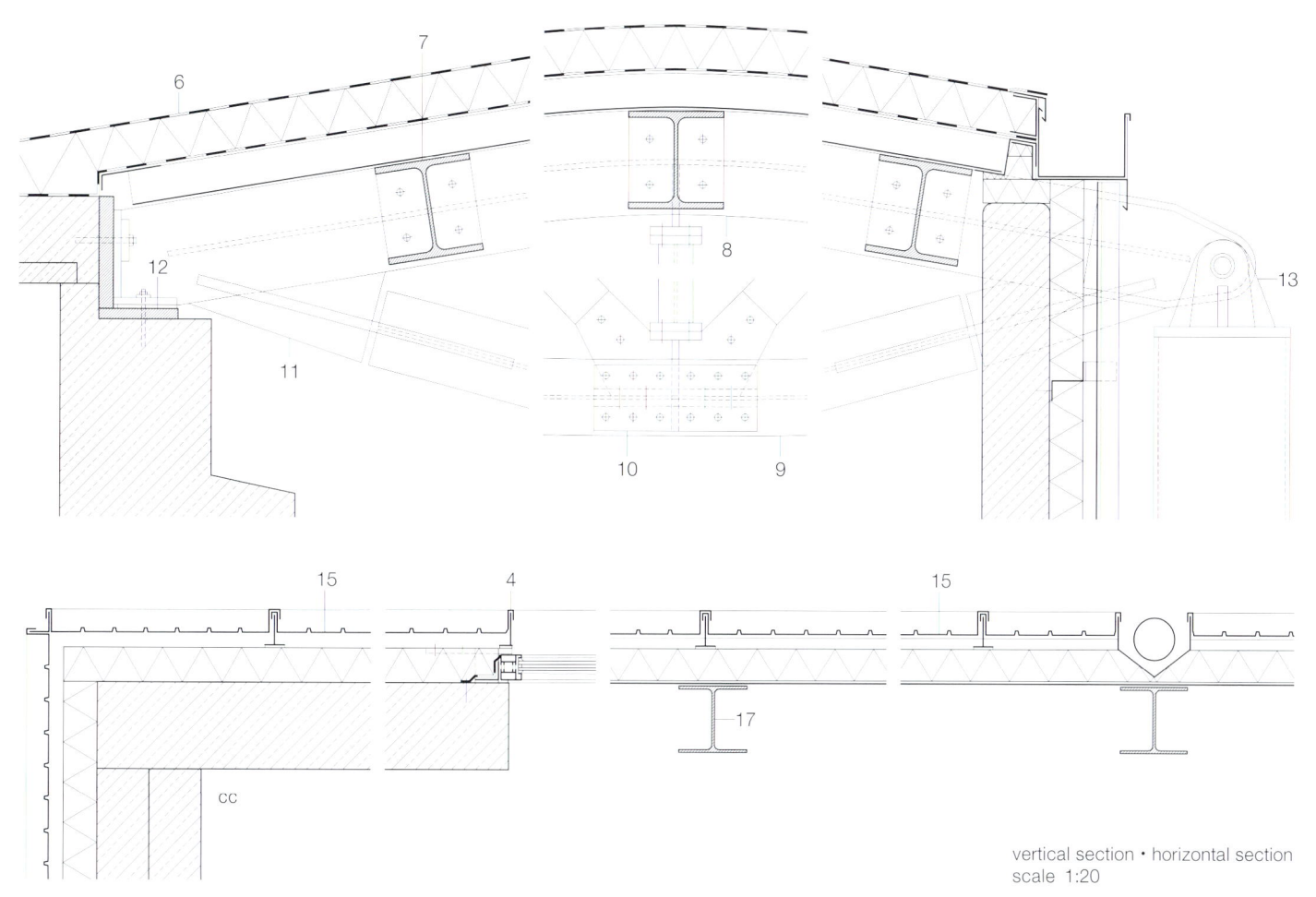

vertical section • horizontal section
scale 1:20

133

"Economy does not mean building cheaply"

Helmut C. Schulitz

How, in your opinion, should the planning process be carried out in order to achieve the most economical solution?
A planning process can only achieve a truly economical solution when the interaction between planning participants is based on mutual respect. It is the overall concept of the building in its complex structure, origin and operation that is important and it is the responsibility of the architect to recognise and reconcile these relationships. Unfortunately, it is often the case that costs dictate the decision-making process, without the client or investor being made fully aware of these relationships, or understanding the difference between cheap and economical construction.

In which particular phase of the planning process, and in what way, were you able to reduce costs?
We were able to save costs in all phases of the process. The potential for cost saving, however, diminishes as the process continues. The initial phase, the project programming, offers the most potential for saving. For example, for the Ice Stadium we worked with a brief which correlated with the directives of the German Ice Hockey League. The calculations, however, resulted in costs exceeding the funds available to the client. Only after intense discussion with investor, operator and the club was it possible to reduce the brief to the absolute minimum, and for brief and budget to coincide.

Which areas, costs and trades offer the largest potential for savings?
Not only the brief, but also the coordination and interaction between the trades, play important roles in the future savings in planning and construction processes. It is often the case that building services are integrated too late in the process. Minimal storey heights may initially appear to contribute to cost savings, but the lack of space for optimal service runs are often the outcome. Not taking into account the space for services when determining the direction and type of structural system also demonstrates the same problem. Of course, it is also possible to make savings within the various trades.

What contributions to cost efficient construction can be made by structural engineers and service consultants?
Structural engineering and technical services offer fundamentally different possibilities for cost savings. This is evident in the technical services of a building; for example, when deciding if air conditioning is to be installed in the gastronomy and office sections of a project. But in almost no other area are the considerations of comfort so closely associated with financial sustainability or running costs as here. We would have been more than happy to provide the stands of the ice stadium with a displacement ventilation system, or to have installed larger video screens. But in consideration of minimal costs only minimal requirements should be fulfilled. Savings in the area of structural planning are really only limited by minimum regulations pertaining to material stability. Considerable savings can, in fact, be achieved by intelligent, creative structural planning. Even initial decisions relating to direction and type of structural system can have great influence. We always develop our structural systems in association with structural planners and other consultants in order to determine the largest degree of savings potential. Unfortunately many architects are incapable of constructional design, or demonstrate minimal interest in it.

How were you able to reduce costs in your project?
It is possible to save costs in almost all trades. Totally or partially prefabricated concrete elements were often implemented, and curved walls were constructed of polygonal beams. It was possible to build the stands out of prefabricated elements based upon existing formwork rather than constructing new formwork. We were able to do without finishing screed in many situations, using power-trowels on the concrete slabs, a standard technique in the USA. Wherever possible we did without suspended ceilings, and the number of lifts was kept to a minimum by locating and fitting them to meet all functional requirements.
The list is endless.

Would you have preferred to use more complicated, expensive solutions in certain positions? Which compromises were necessary?
It was obvious to all participants that the original design would have been the better solution (see perspective); unfortunately it wasn't covered by the budget. In the revised version, we would have all preferred to have more funds available. For example, the originally planned stair at the main entrance would have been attractive, and we would have preferred to have provided more comfortable seating for the VIPs, most of whom are older. The decision to install affordable, cloth-covered shell seats, however, also signalised the omission of social differences. The continued demolition of the old hall would have been preferable; the renovation demanded increased efforts in planning and was also responsible for constructional and aesthetic drawbacks. The solutions to all compromises, however, were so convincing that they do not give the impression of being compromises at all.

Were more expensive solutions chosen in particular areas for which it was necessary to make savings in other places?
No, the budget didn't allow that.

What exactly are the economical advantages of the building materials you selected compared with other building materials?
The advantages of more economical building elements and materials can be seen in the aesthetics of simplicity. Expensive materials are not always more attractive. Sometimes, however, the only advantages are the prices; for example, it wasn't important to us whether the roof structure was executed in timber, or a composite structure of timber and steel. We planned and documented both alternatives simultaneously and the more economical solution – in this case, the steel structure – was finally selected. Sometimes, of course, more economical building elements bring disadvantages with them. We were forced to accept standard format pre-cast elements rather than the desired customised elements which would have fit our formal concept better but were simply not in our budget. These disadvantages also demanded additional work.

How do you imagine the future of cost and resource efficient construction to be?
The future of cost and resource efficient construction is certainly contradictory. In view of the fact that the world's resources are limited, and that more people live in poverty than wealth, our responsibilities lie in the development of a maximum of living space with a minimum of material and financial assets. Cost efficiency and the conservation of resources are not always identical in construction. With the increase of salaries in the industrial nations, cost efficient building is often perverted by accepting an increase in materials in order to reduce salaries and even planning costs. All planners are aware that cost efficient building actually requires more planning, but this is not compensated by the clients. More and more investors and clients are of the opinion that by cutting planning fees, costs will be reduced. For example I can understand when a budget for structural engineering is so severely reduced that the engineer is unable to calculate an economical design; when he only calculates the critical beams and columns and then applies these elements to the entire project. The material waste in such situations is often so great, and so expensive that the savings made on fees are greatly outweighed by the increase in material costs. This is usually unrecognised by to the majority of clients. The result of all these cost saving measures is that mass-produced buildings are created, as has been known for decades in the USA as "pre-engineered buildings".

We live in schizophrenic times: on one side there are cheap constructions which clearly demonstrate that costs have been saved in both planning and construction, on the other side buildings are created where costs seem to have absolutely no bearing, in the manner of the "Bilbao effect". When it is anticipated that accrued additional construction costs – 304 million Euros for the completed concert hall in Valencia compared with the originally quoted 84 million – can be absorbed by increasing tourism profits (as in Bilbao), then even these types of buildings can be considered cost efficient in a capitalistic sense.

Exactly what cost efficiency means, is a point of controversy. One thing is, however, certain; no architect ever admits to building in a non-efficient manner. There is no area of construction where so much incorrect information is given as in costing. Finally, cost planning allows such a huge range of flexibility and is so difficult to control due to the fact that no one truly understands, manages or takes responsibility for the complex constructional interactions involved. For example, no jury member in an architectural competition is ever held responsible for his decision to select a particular scheme demonstrating a complicated design, which was not fully considered in all its aspects and subsequently leads to exploding costs and extended construction times. The various parties involved in construction concentrate on limiting their own responsibilities. Structural planners and consulting engineers are as unwilling to be involved in the aesthetic design of the building as the architect is in the production of the building. Expecting cost efficiency to be controlled by the project manager is doomed to failure as a result of his professional training which has little or nothing to do with design.

When cost and resource efficient construction is to have a future, the concept of professional capacity must remain a fundamental element of the design process. This must not, however, be restricted to functional performance alone, but should include all elements of the building; the construction, technical services, orientation to the sun and the resultant solar, ventilation and physical form of the building. The goal should be to fulfil all demands of the brief while simultaneously minimising the amount of materials and energy used in both the process and the product.

Often the failure to produce a cost efficient construction is not the fault of the planners. Many clients are at fault due to their particular prejudices. Not every simple form is economical and not every complicated form is expensive. Thirty years ago our office in Los Angeles (U.I.G.) was required to simplify the facade of a project, which was exactly on budget, so that the building would not appear too expensive because the client wanted to avoid any possible criticism. We experienced a similar situation, but with different result, with an ice skating, training and ball complex in Germany. We designed a dynamic form based upon the volume which coincided exactly with the competition brief and a structural system based upon the moment curve. We won the competition and were awarded the contract after an independent consultant proved that the design was able to be realised on the given budget. The client considered the building form to be too complicated and was of the opinion that the desired project expansion of 10% could be absorbed by simplifying the form and architectural design. After we stated that this policy was unrealistic, it was decided that another competition entry should be built. The simpler, box-shaped design was guaranteed for a total cost of 22 million Euros. The result of this theoretically cost efficient decision was that to date the building is still uncompleted and has already exceeded the given budget by 8 million Euros. The prejudice that simpler forms are more cost efficient led to cost increases greater than the total costs of our design for the Ice Stadium in Wolfsburg.

Service Centre in Frankfurt am Main

Architects: Dietz Joppien, Frankfurt am Main

Light-weight concrete single-layer external walls
Reduced fit-out for diverse user groups

Nicknamed the UFO, this complex which accommodates office and event facilities is located in Frankfurt's rather disorganised and erratically developed eastern industrial area. The introverted, five-storey construction, with triangular footprint, deliberately distances itself from its neighbours. A rigid grid has been applied to the street frontage, manifested as single-format window perforations. Only the corner of the building, at the street intersection, opens up to a height of eight metres, enticing observers with views of the glowing red entrance ramp. The 42-metre-wide curved opening is clear of columns and supports the upper facade, which itself is reminiscent of a "Vierendeel" truss. From the ramp one moves towards the entrance into the event-area located on the ground floor, while another entrance leads to the upper storey where the office spaces are organised around the triangular courtyard. Access walkways interconnect the retail zones, which are of varying sizes and are let with nothing other than the service cores as facilities. All installation ducts are visible, which enables any desired alterations to be both speedy and economical. Internal subdivisions and material selection are determined by the individual tenants. The demands of a wide variety of users; from retail, trade and service industries, can therefore be met. All storeys are designed for large applied loads with generous ceiling heights (3.67 metres) and wide column spacings (14.36 metres). The load-bearing internal walls and the internal faces of the peripheral walls are left as exposed concrete.

Emphasising the monolithic appearance of the building are the single-layer external walls; they are constructed of 50-cm-deep light-weight concrete and satisfy thermal requirements without the need for additional insulation. There are no expansion joints, a minimum of tie holes, and all concreting joints and boxed openings were specified by the architects prior to construction. Each phase of the concreting was carried out in direct contact with the preceding phase, so that the intermediate joints receded and an uninterrupted, continuous expanse was achieved. To ensure uniform setting of the concrete, it was temporarily protected by a spray-on membrane for a period of three weeks. The membrane disintegrates under ultra-violet light. Consistent savings in fit-out and the application of a single window format absorbed the costs of the more complex facade construction. The facade takes on a representational role and enhances the perceived value of the project.

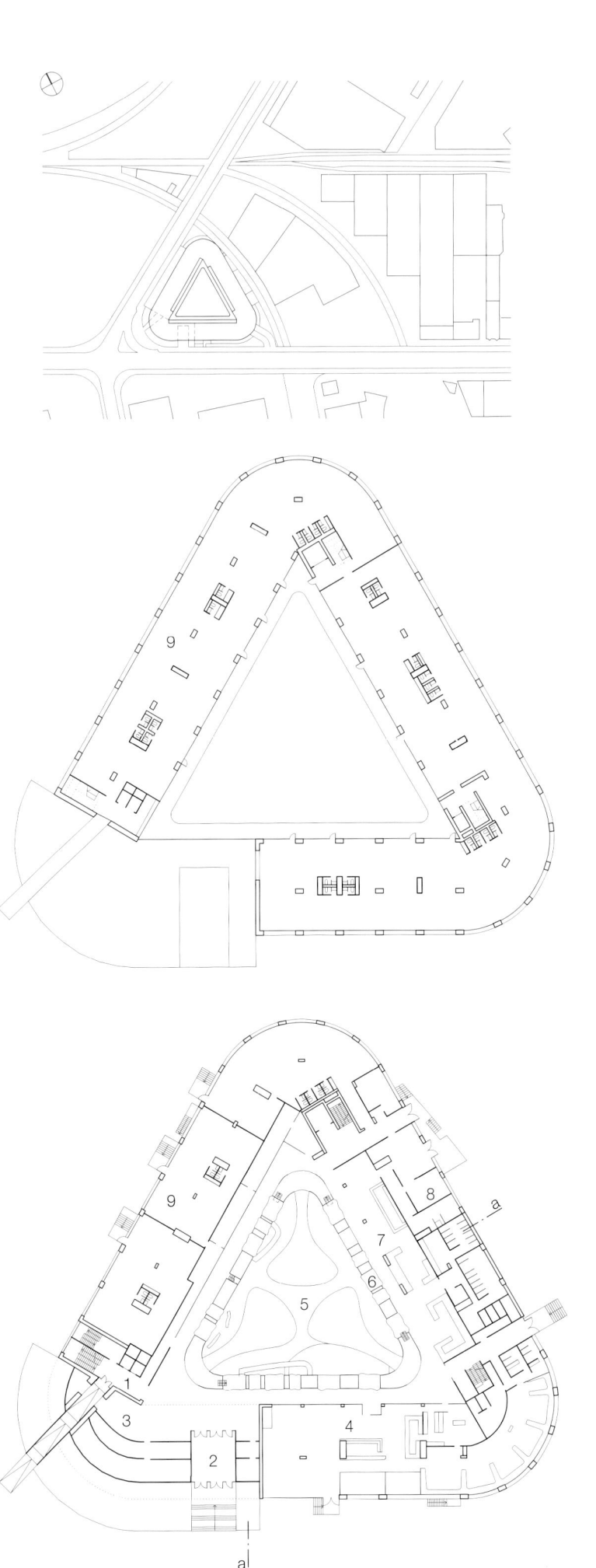

ground floor plan
first floor plan
section
scale 1:1000

1 entrance to first floor
2 entrance to ground floor/
 basement
3 cloakroom
4 restaurant
5 dance floor
6 niche
7 bar
8 office
9 function unit/retail space

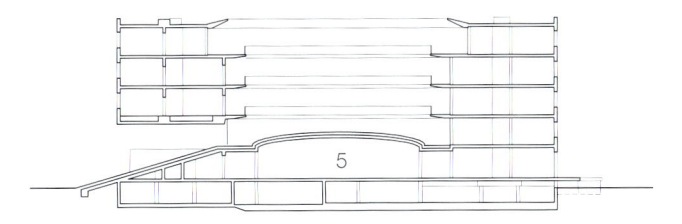

aa

Project details:
Usage:	retail and service facilities
Access:	open walkways
Construction:	reinforced and light-weight concrete
Internal ceiling height:	3.6 m
Construction cost (net):	14.6 million Euros
Cost per m²:	1,097 Euros
Total internal volume:	88,000 m³
Total built area:	21,749 m²
Total usable area:	13,300 m²
Total site area:	5,972 m²
Construction period:	18 months
Date of completition:	February 2004

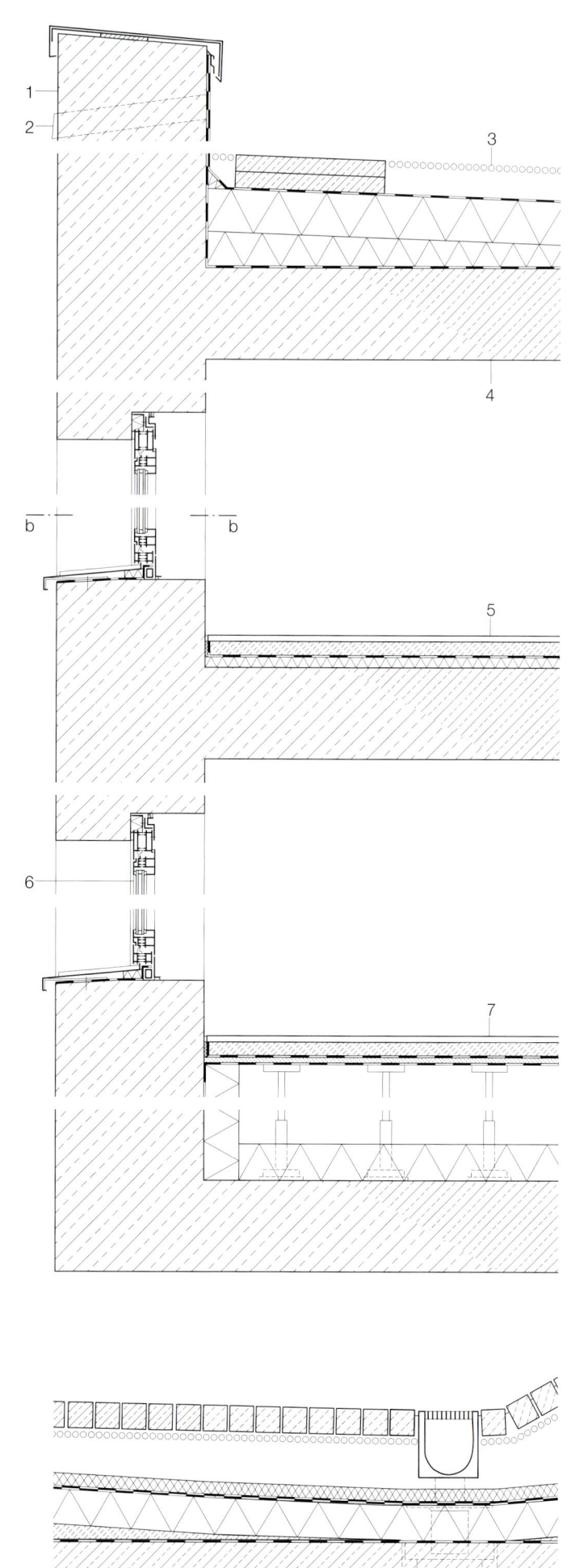

138

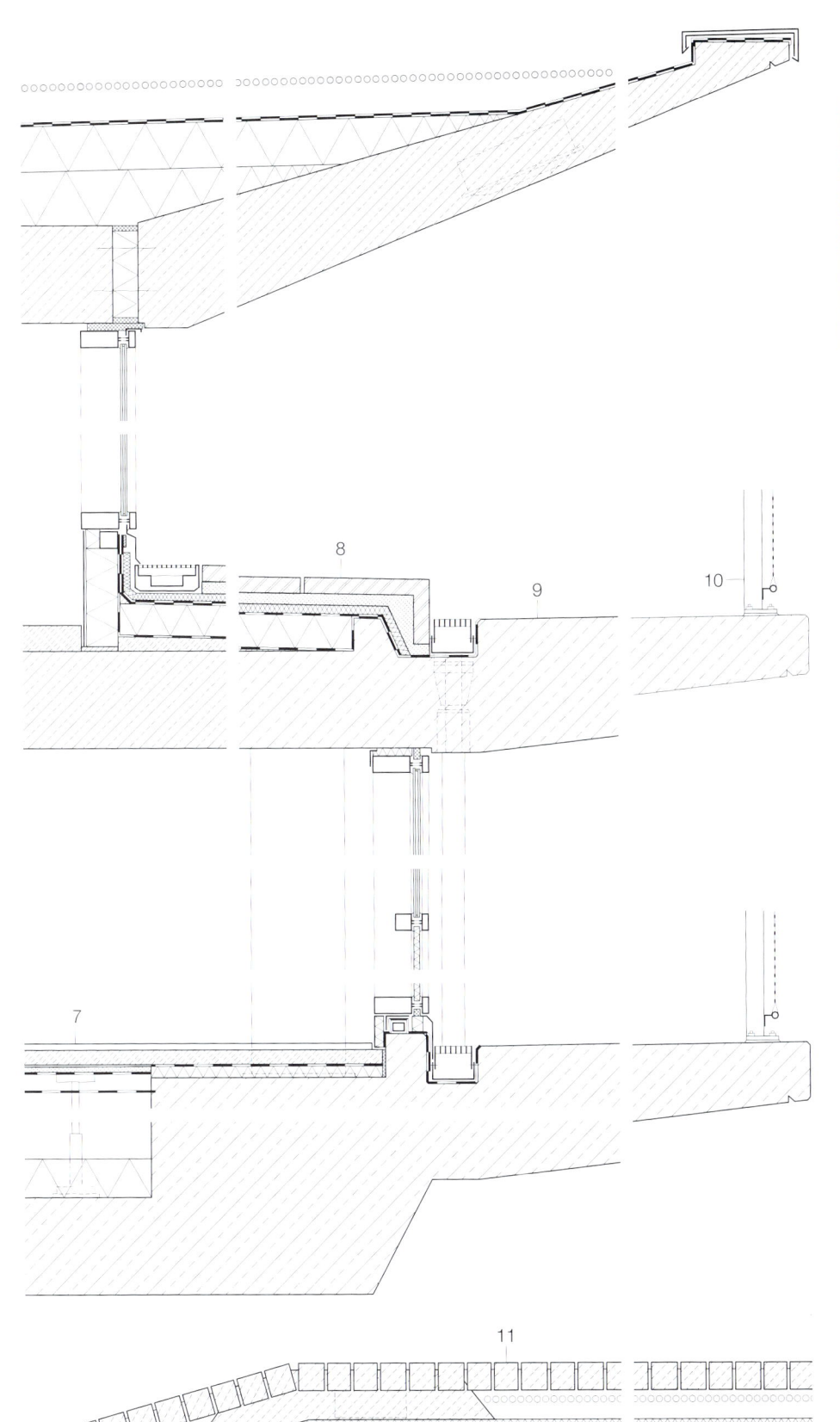

horizontal section
vertical sections
scale 1:20

 1 500 mm light-weight concrete
 2 overflow
 3 100 mm gravel
   double-layer bituminous sheeting
   140 mm thermal insulation
   insulation, with falls
   vapour barrier
   300 mm light-weight/reinforced concrete
 4 light-weight/reinforced concrete,
   with continuous poured connection
 5 30 mm various floor finishes
   45 mm screed, separation layer
   35 mm impact-sound insulation
   300 mm light-weight/reinforced concrete
 6 double-glazing in steel window frame
 7 30 mm various floor finishes
   40 mm screed, separation layer
   18 mm cavity floor slab, vapour barrier
   floor slab sub-construction
   120 mm thermal insulation
   300 mm light-weight/reinforced concrete
 8 optional partition wall anchor
 9 400 × 400 × 50 mm anthracite walkway pavers
   35 mm mortar bed
   20 mm drainage mat
   double-layer sealing layer
   100 mm thermal insulation
   300 mm light-weight/reinforced concrete
10 access walkway: waterproof light-weight concrete
   coating, epoxy-resin sealant
11 balustrade posts:
   60 × 75 mm metal sheeting, bent to form,
   welded to base plate and galvanised
   intermediate tensioned steel cables
12 80 mm orange pigmented concrete pavers
   90 mm sand bed, 40 mm drainage layer
   20 mm multi-layer polymer-bituminous
   membrane, 120 mm insulation
   vapour barrier, 30 mm reinforced concrete

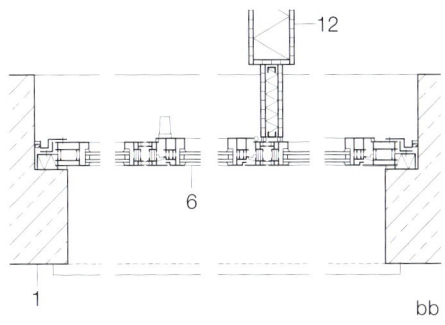

bb

139

Planning as „refined carcass" construction: economical, functionally neutral, robust

Matthias Schönau

The concept underlying the U.F.O loft and commercial development in Ferchenheim, Frankfurt am Main (which won this year's Detail Prize) was for a building "without a specific programme". A better description of the goal would perhaps be "programmatic flexibility", in which stress is laid on providing for a large number of options. This concept was the outcome of two constraints: the economic need to develop the site within a certain time; and the impossibility of getting potential leaseholders to commit themselves for the whole period between the application for building permission and the completion of the scheme. It was thus decided that there should be scope to accommodate not just one specific function, but all conceivable uses. Situated in a classical "industrial location" in the east of Frankfurt, the building was also to activate a process of transformation in a hitherto neglected area. The sculptural volume that was realised, distinguished by its simple, clear-cut geometry, makes the requisite impact and also creates a sense of identity. In addition, the material chosen – exposed concrete – allows a balance to be achieved between physical robustness and haptic refinement, a balance that is necessary if the building is to assert itself in its "raw" surroundings and to generate greater architectural quality in the area. In addition to its urban-planning function, the structure was to comply with the spatial and technical needs of as broad a range of users as possible, who were completely unknown at that time.

The concept therefore aimed to achieve a maximum of structural adaptability in terms of function, spatial dimensions, load-bearing behaviour, access and circulation, technical services and finishes.

Low-cost construction
In planning low-cost construction, the life span of a building also has to be taken into account. Studies of older commercial or industrial models show the longevity and adaptability of functionally neutral buildings with simple load-bearing structures. Creating a robust design was, therefore, a central aspect from the start. Allowance had to be made for frequent conversions. The requisite programmatic latitude lay in the various systems – structure, technical installations, finishes, circulation, and layout geometry.

Access and geometry
Within the triangular layout of the building, access is systematically organized via three vertical cores. At the same time, the location of the main entrance, and the goods lifts for deliveries at the rear, define the address and external orientation. The cores are linked by courtyard access galleries on each floor. Along these routes, the leasable areas are reached on every second axis through the transparent courtyard facade. This "excess of access", as it might be described, facilitates a free division of the floor area into leasable units of different sizes. With a width of 2.5 m, the access galleries are larger than necessary for escape routes or normal circulation functions. They offer certain urban qualities, as the seating groups, pot plants and grills installed by tenants demonstrate.

Structure
Adequate provision for live loads (5–10 kN/m^2, even in the access galleries) means that all upper storeys are suitable for a variety of potential commercial functions. Large spans and generous ceiling heights, a restricted number of columns, and flat soffits also ensure more latitude in the finishes.

Technical fit-out
Only with the provision of adequately dimensioned shafts (relatively closely spaced and with ventilation as an option) was it possible to guarantee a truly free division into leasable units of different sizes. It was necessary to ensure that alterations to the installations could be simply made at a later date, whereby the construction has to comply with a requirement for 1-hour fire resistance. In view of the ceiling heights and the depths of lintels over the windows, suspended soffits are also conceivable.

Interface between carcass and fit-out
The building was planned as a "refined carcass structure". In other words, the public areas and circulation routes were finished to a normal standard, whereas the finishes to the leasable areas extended to no more than the flooring and screeds, shafts with basic service connections and electrical distribution points. Toilet connections are prepared within the shafts, but the WCs themselves have to be installed by the tenants. The over-dimensioning that occurred and the provision of a greater density of installations than was actually required, in order to cope with all eventualities, meant a higher level of investment, so that the development could not at first sight be regarded as low-cost. Nevertheless, the greater adaptability and leasability resulting from these measures mean a much longer life for the building, so that they, too, are ultimately of economic advantage.

Concrete as a material

We saw concrete as the ideal material to meet the requirements of the brief in an economical form. Acoustics, fire-protection, thermal insulation and load-bearing ability were not to be segmented into specialized additive functions, but united in a synthesis of design and materials. To achieve a monolithic form, in-situ concrete was specified. The risks involved in executing an exposed concrete facade entirely in-situ were accepted by the client in order to achieve a bold architectural expression. The risks, however, were reduced by the self-critical approach of the main contractor, and precise conditions were drawn up in regard to quality control.

The 50 cm load-bearing external walls, with 40 per cent of their area taken up by window openings, were a major factor in creating the impression of a solid structure. Like the access galleries, the walls were constructed entirely in lightweight concrete. The requisite quality control of additives and the production of the material set very high demands for planners and suppliers alike, since acceptance of adequate thermal insulation on the basis of physical data (specific gravity) was granted only as an exception and subject to empirical proof. In other words, no changes at all were allowed in the mix. Many decisions relating to details were subordinated to the central goal of achieving a monolithic appearance. The various sections of the concrete were simply abutted to each other; and anchor holes were avoided as far as possible. Because of the consistent construction grid, a high degree of repetition was possible in the formwork; but a seamless link between CAD drawings and the creation of formwork on a CAM-supported basis was not possible to the desired degree in geometrically sensitive areas such as the entrance and the dome.

The concept of a rough concrete building is rounded off architecturally by a high degree of design input in the public areas; e.g. the entrances, staircases and access galleries. The lack of details and the monochrome nature of the bulk of the structure are complemented here by the economically detailed design and selection of materials, consisting of stone, wood and steel mesh.

The significance of using a single material, namely reinforced concrete, can be summarised as follows.

- The architectural concept of a large-scale, expressive monolithic form could be implemented with exceptional consistency.
- The planning and execution required greater elaboration only in special situations with more complex geometry (e.g. projecting elements, the dome and entrance).
- Planning exposed concrete surfaces is always a complex matter. The reduction of details for finishes and junctions made itself felt in the course of the planning.
- Designing a solid structure with few details greatly improved the flexibility of the building in terms of alterations.
- The use of lightweight concrete was far more demanding when it came to maintaining the requisite quality.
- The costs of lightweight concrete are markedly higher than those of normal reinforced concrete because of the more elaborate production process and the use of special aggregates (expanded clay or shale). The cost of the lightweight concrete facade as executed was only slightly less than that of a simple structure with full thermal protection. In the latter case, however, it would not have been possible to realise details such as the broad cantilever over the entrance.

Conclusion

Achieving a synthesis of the architectural and structural concepts on a lasting, economical basis was possible only in the present uncompromising form. In the meantime, the building has become a landmark, and the investment in flexibility has proved its worth.

1 Interior of the U.F.O. loft and retail building, Frankfurt am Main

Office and Training Centre in Dresden

Architects: Heinle, Wischer and Partner, Dresden

Modular systems with prefabricated elements
Sustainable energy savings

The architects set themselves the ambitious goal of designing a low-budget building, compatible with its heterogeneous urban environment, but one that was still capable of satisfying higher demands of quality than those of the existing commercial developments. They appear to have been successful with this well proportioned building located on the outskirts of Dresden. The regular construction grid and the calm cuboid form introduce a sense of order to the chaotic surroundings. The organisation for trade and commerce in Saxony needed new administrative and training facilities. This L-shaped building effectively creates space and volume while the facades are clearly articulated and precisely detailed. A three-storey foyer links the various functions and provides space for public and representative functions, while the administration benefits from an introverted, glazed atrium.

Cost-saving strategies

To reduce costs, the architects restricted the range of materials and unit sizes. Prefabricated multiplex aluminium window elements were fitted into the modular system, while the concrete structure between is clad with black fibre-cement panels. All details were designed to achieve a flush finish. The construction is based on the application of standard products and dimensions, enabling simple implementation by a large number of firms. This also caused highly competitive tenders to be received. Parallel planning and detailing processes allowed for flexibility of response and adaptation to immediate requirements on site.

The central atrium serves the needs of the passive indoor climate and ventilation concept; allowing solar thermal gains to be made in winter and assisting the ventilation system throughout the whole year. Preheated fresh air is drawn in through a sub-floor duct. The extracted air escapes via openings in the glazed roof, while fresh air flows in replacing it.

Project details:
Usage: administration building
Construction: reinforced concrete
Internal ceiling height: 2.63 m
Total cost of construction: 2.8 million Euros
Cost per m² area: 970 Euros
Total internal volume: 8,556 m³
Total floor area: 2,881 m²
Usable area: 1,465 m²
Site area: 7,600 m²
Built site area: 1,042 m²
Period of construction: 15 months
Date of construction: 2002

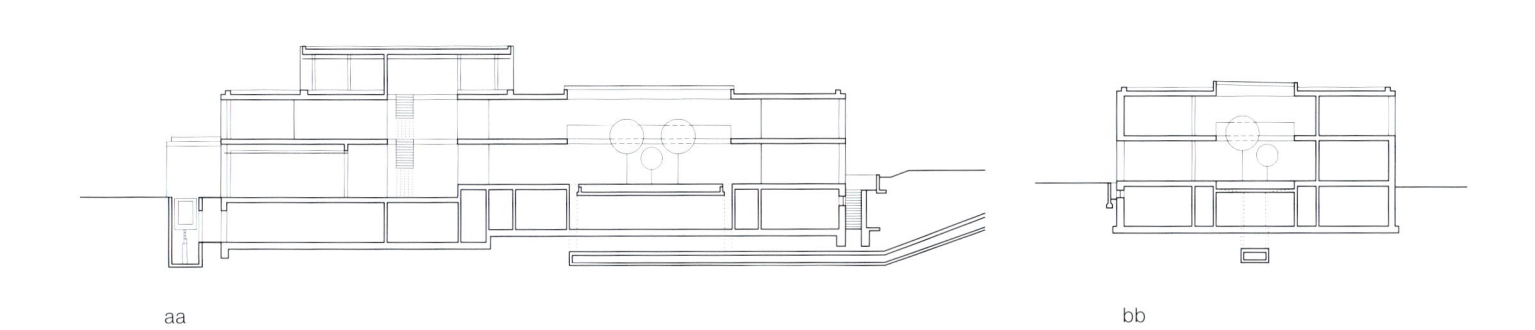

aa

bb

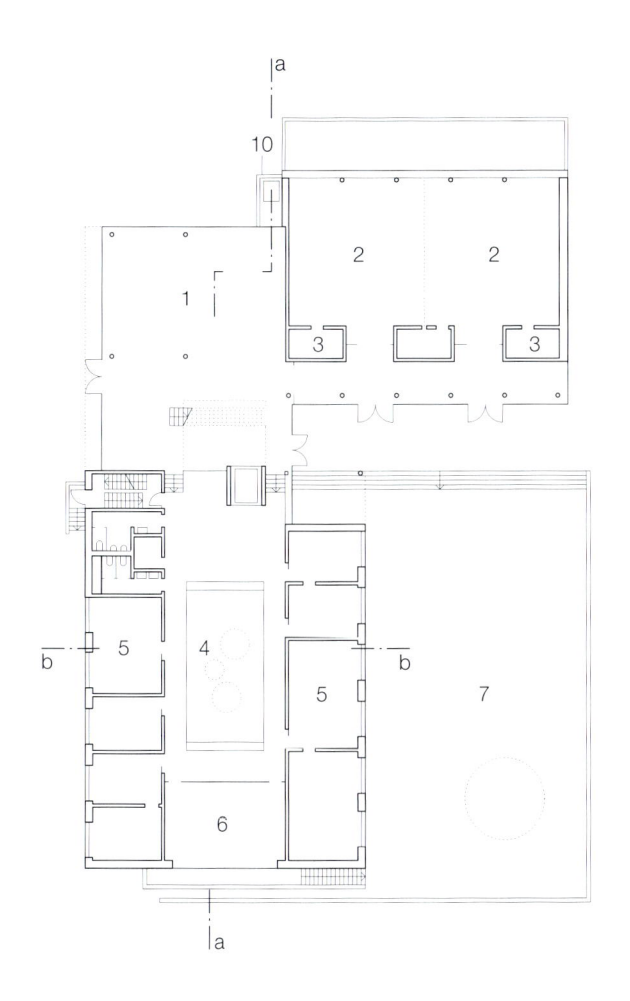

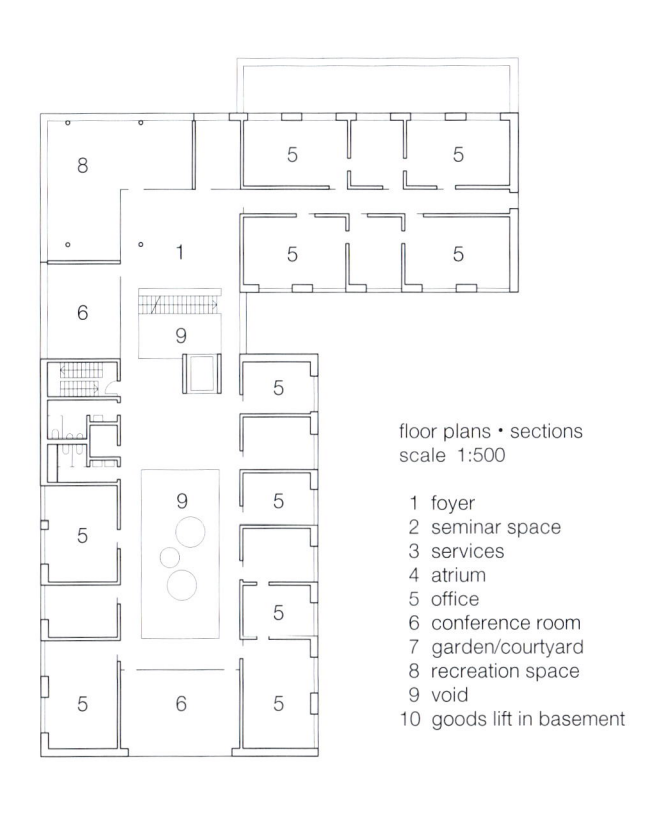

floor plans · sections
scale 1:500

1 foyer
2 seminar space
3 services
4 atrium
5 office
6 conference room
7 garden/courtyard
8 recreation space
9 void
10 goods lift in basement

143

vertical section through atrium
sections through skylight
scale 1:20

1. roof construction:
   80 mm bed of gravel, protective mat
   2.3 mm self-adhesive plastic sealing membrane
   1.8 mm bituminous-rubber covering with glass-fibre inlay on
   160 mm polystyrene thermal insulation
   1.2 mm vapour barrier with glass-fibre reinforcement
   220 mm reinforced concrete slab
2. birch laminated construction board element with aluminium bottom-hung panel with double glazing:
   2× 4 mm laminated safety glass +
   25 mm cavity with aluminium louvers +
   2× 4 mm laminated safety glass
3. office floor construction:
   2.3 mm textile flooring
   55 mm screed with synthetic fibres
   0.2 mm polythene separation layer
   20 mm mineral-fibre impact-sound insulation
   40 mm polystyrene insulation
   220 mm reinforced concrete slab, render
4. 8 mm fibre-cement sheet, rivet fixed
5. atrium floor construction:
   100 mm river gravel on geo-textile membrane
   300 mm intensive substrate layer
   1.5 mm filter mat
   120 mm drainage layer
   3 mm protective layer
   1.5 mm root-resistant layer
   3 mm protective layer
   220 mm reinforced concrete floor slab
6. roof construction:
   100 mm extensive roof planting layer
   sedum shoots
   mineral substrate layer
   system filter, drainage panel, separating mat
   self-adhesive plastic sealing membrane
   1.8 mm bituminous-rubber covering
   160 mm polystyrene thermal insulation
   1.2 mm vapour barrier with
   glass-fibre reinforcement
   220 mm reinforced concrete slab
7. double glazing, designed to bear foot traffic:
   2× 5 mm laminated safety glass +
   16 mm cavity + 10 mm toughened glass
   30 × 50 mm aluminium facade section welded to steel section
8. 50 × 170 mm aluminium RHS
9. 20 × 170 mm steel T-section, painted
10. double glazing, not designed to bear foot traffic

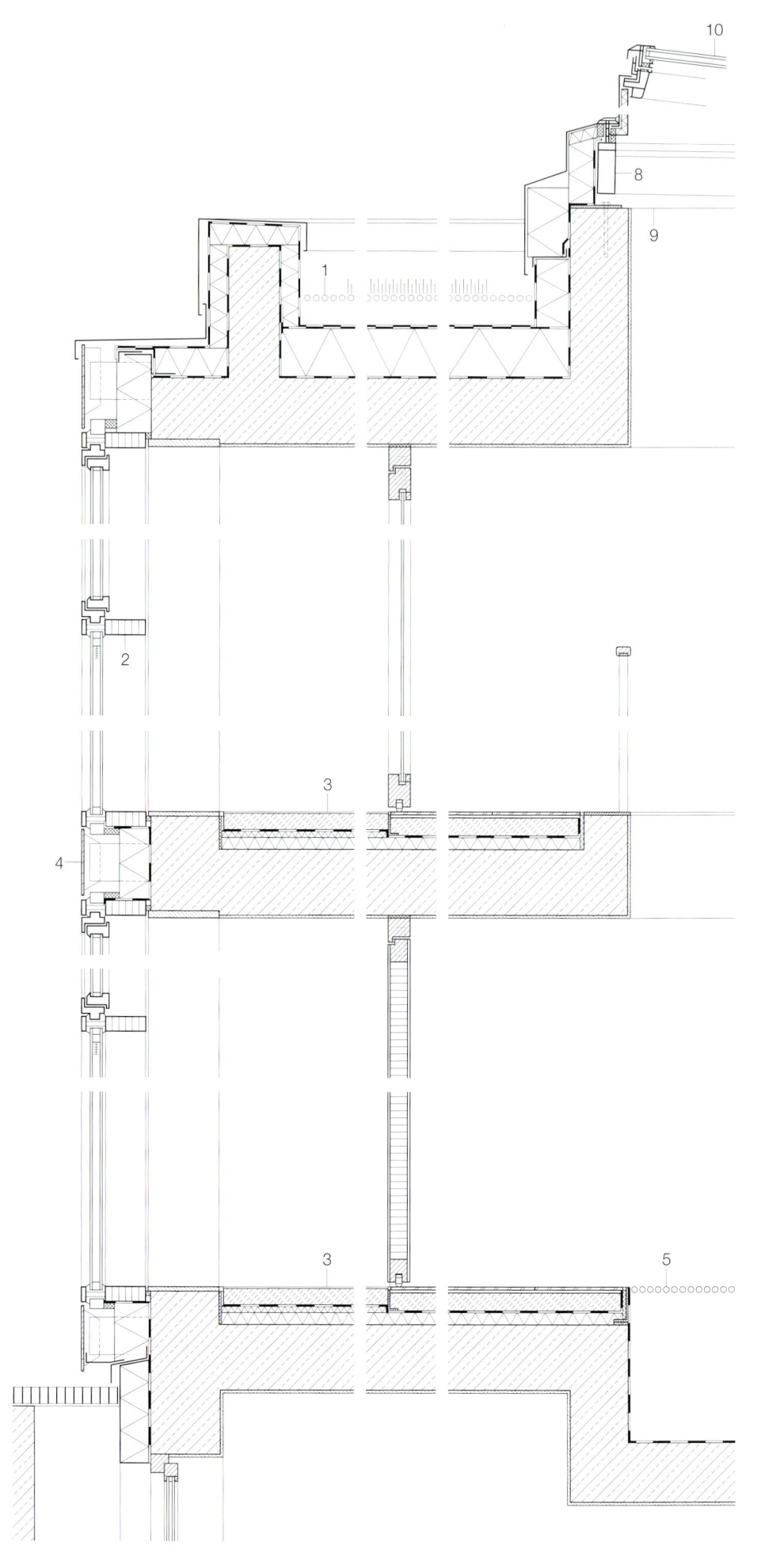

145

Production Building in Grosshöflein

Architects: querkraft architekten, Vienna

Individual spatial concept
Creative application of standardised products

How does a company that prints large-scale banners for cultural events and advertising purposes present itself to the world? The answer in this case is certainly not a spontaneous, spur of the moment idea, but rather the product of intense and long-term discussions with the client.
For the new production building, the printer deliberately chose a location immediately adjacent to an autobahn. At night, coming from Vienna shortly before Eisenstadt, one becomes aware of the rear-lit image of a mountain panorama. Located in a low-lying, sparsely populated rural region preceding the eastern foothills of the Alps, it is unaffected by other light sources. Branching off the highway to gain a closer view of this phenomenon and following a number of smaller roads, one finally approaches the free-standing structure. The rear face of the building is emblazoned with the company's 60-metre-long slogan; "unignorable". The building concept is not only confined to the special effects of these two representational facades. Its many spatial qualities serve to improve both the working environment for the employees and the effectiveness of the production process.
The office spaces and printing shop are located on different levels yet maintain direct visual contact. The printing shop encompasses the entire northern side of the ground floor level of the premises. To separate the various production areas functionally and acoustically, without impairing the sense of spaciousness, a transparent plastic membrane was drawn across the width of the hall. Various storage rooms and a metal-working shop are located within the building on the southern side, and comprise approximately half of the building's width. An open office space is located above this fire-resistant reinforced concrete cube, while a broad walkway between the offices and production hall allows long-distance views of the large scale printing operations.
In order to remain within the tight budget, standard products were selected which could be further developed by the architects, in conjunction with the contracting firms. For example, the width of the structural grid was not fixed prior to the tender period, because different systems are based upon different dimensions. Technical services were also structured upon economic solutions; service ducts were located so as to provide a walkway for the employees.

Project details:
Usage: Production building
Access: open walkway
Internal ceiling height: 3.77 m offices
7.41 m production
Construction type: reinforced concrete, steel frame
Total internal volume: 12,989 m³
Usable area: 2,341 m²
Site area: 5,070 m²
Built area: 1,672 m²
Construction cost: 1.09 million Euros
Cost per m² usable area: 466 Euros
Period of construction: Dec 2001–June 2002

floor plans · sections
scale 1:750

1 parking
2 store
3 printing shop
4 manufacturing
5 car rental
6 entrance
7 reception
8 staff offices
9 management
10 staff room

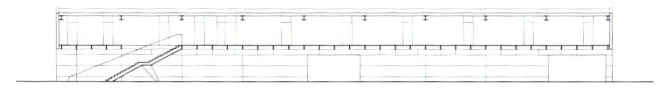

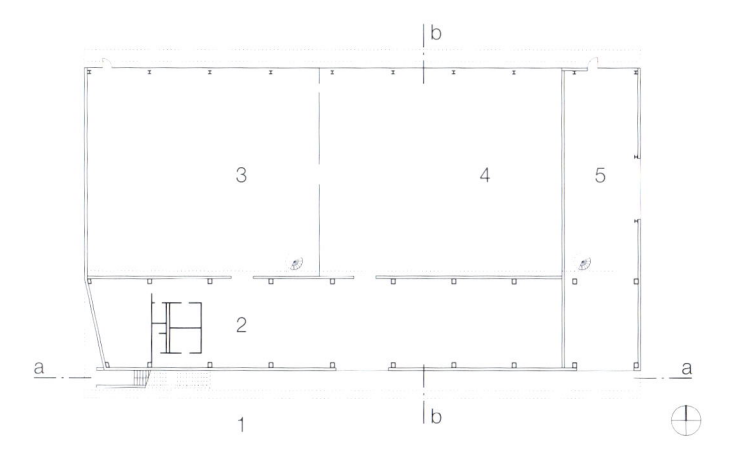

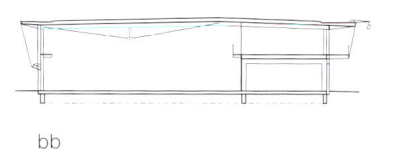

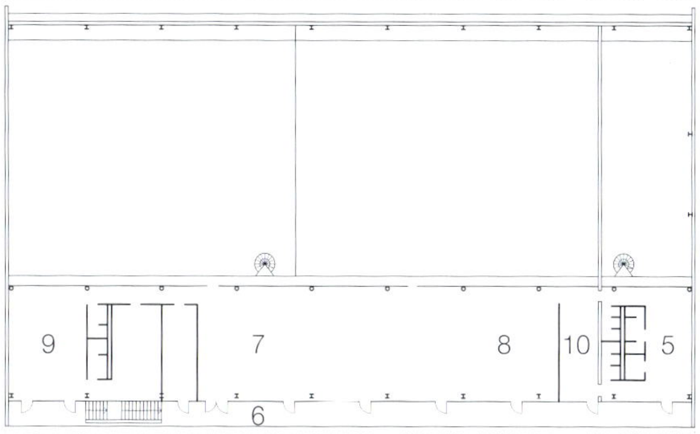

148

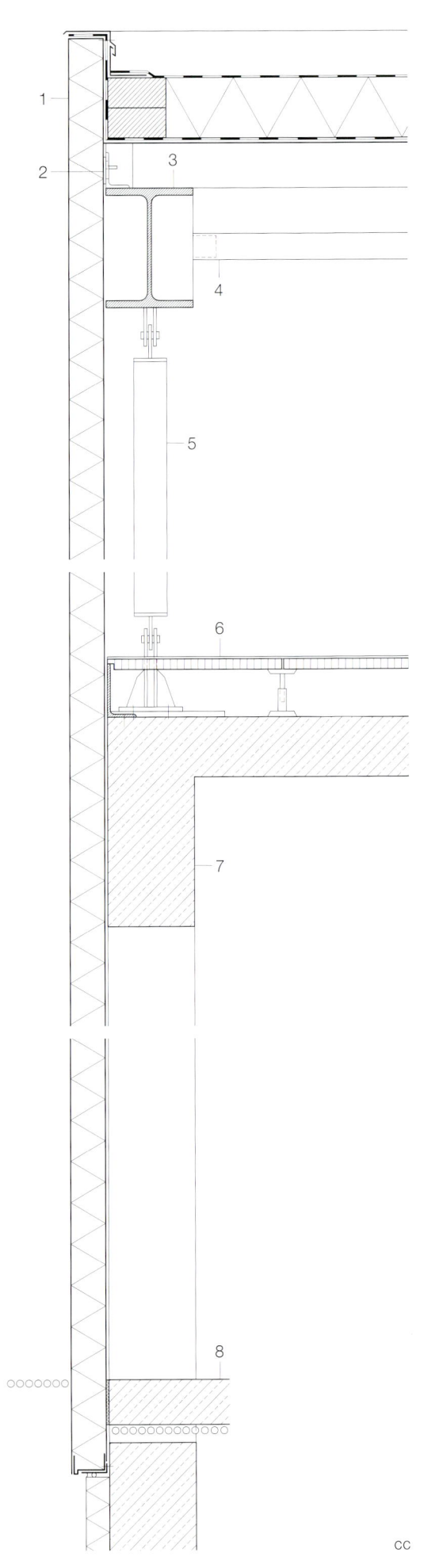

detail section
scale 1:20

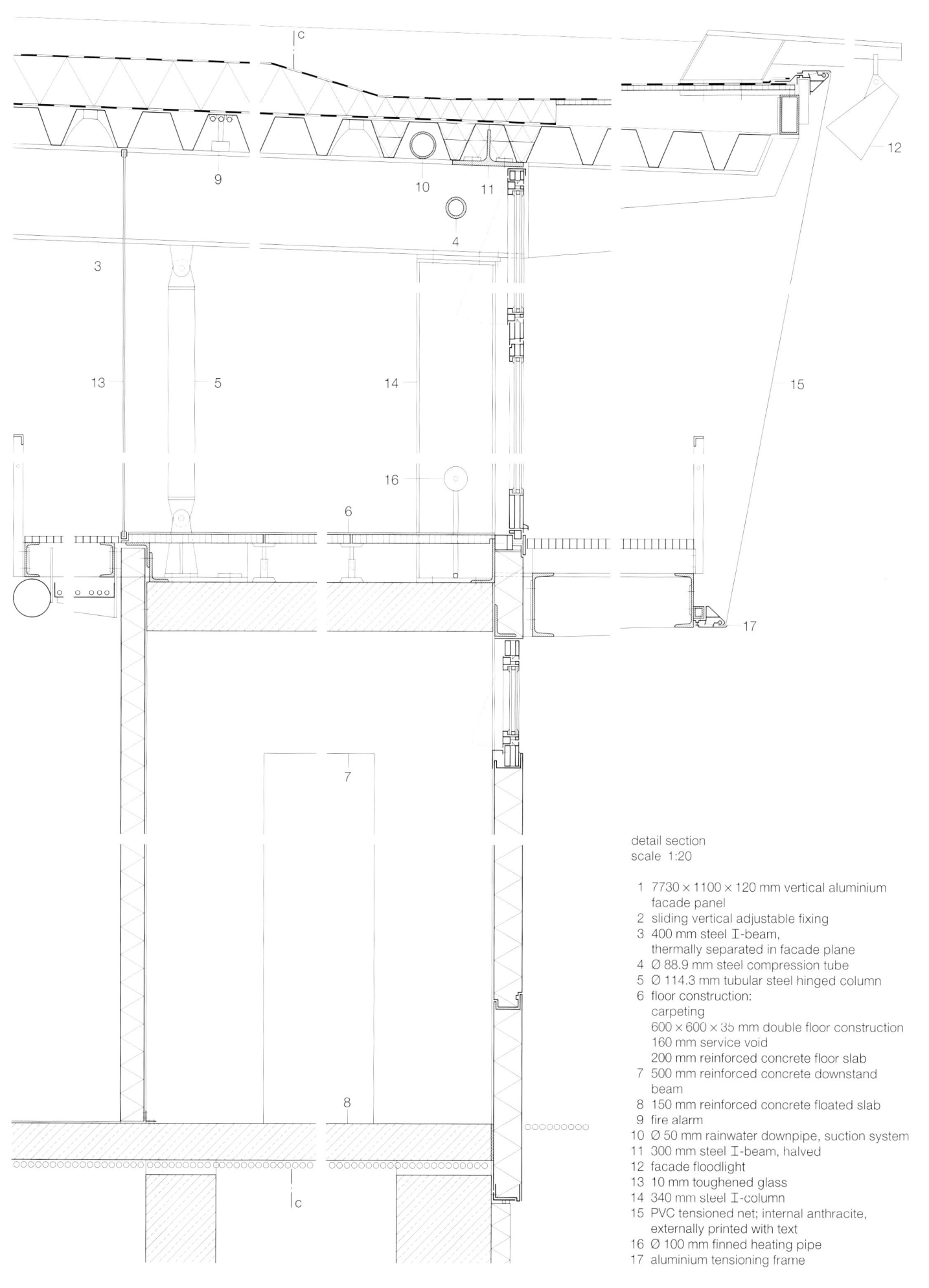

1. 7730 × 1100 × 120 mm vertical aluminium facade panel
2. sliding vertical adjustable fixing
3. 400 mm steel I-beam, thermally separated in facade plane
4. Ø 88.9 mm steel compression tube
5. Ø 114.3 mm tubular steel hinged column
6. floor construction:
   carpeting
   600 × 600 × 35 mm double floor construction
   160 mm service void
   200 mm reinforced concrete floor slab
7. 500 mm reinforced concrete downstand beam
8. 150 mm reinforced concrete floated slab
9. fire alarm
10. Ø 50 mm rainwater downpipe, suction system
11. 300 mm steel I-beam, halved
12. facade floodlight
13. 10 mm toughened glass
14. 340 mm steel I-column
15. PVC tensioned net; internal anthracite, externally printed with text
16. Ø 100 mm finned heating pipe
17. aluminium tensioning frame

149

sections   scale 1:20

1 stretched screen, externally printed
2 aluminium tensioning frame
3 halogen floodlight
4 white reflecting film, 100 mm thermal insulation PE membrane, 153 mm trapezoidal sheeting
5 membrane, 18 mm particle board, steel sections, 150 × 280 × 0.7 mm trapezoidal sheeting
6 membrane, 100–200 mm thermal insulation, trapezoidal sheeting
7 Ø 88.9 mm steel compression tube
8 400 mm steel I-beam, trussed on underside
9 300 mm steel I-column
10 bracket for subsequent addition of catwalk
11 cable duct
12 Ø 100 mm finned heating pipe
13 4 mm PVC flooring, 150 mm reinforced concrete floated slab on recycled concrete sub-base

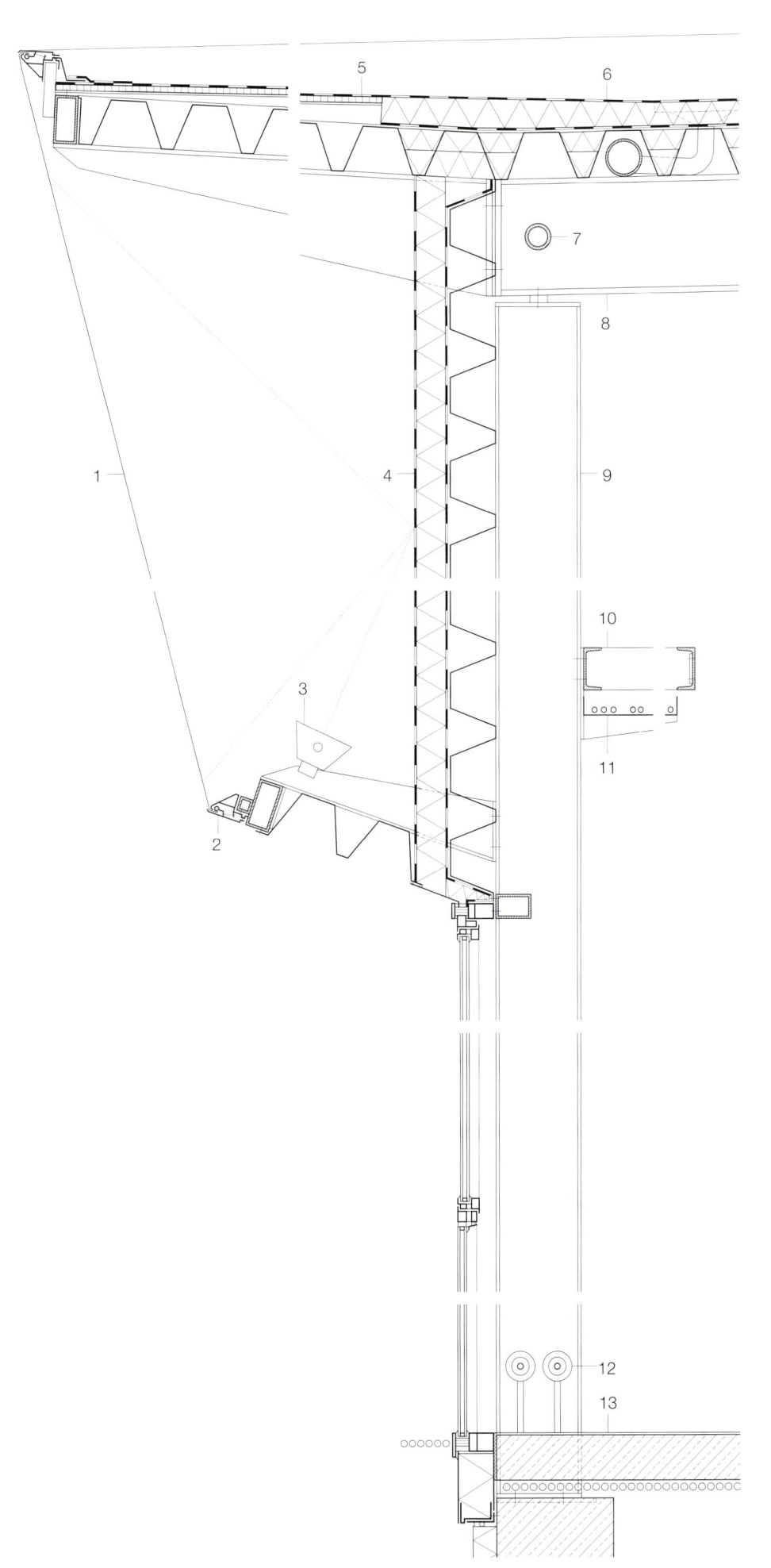

Cost Management

querkraft architekten with Erwin Stättner
Erwin Stättner, a freelance assistant with querkraft architects, was the project architect for the "Trevision" scheme.

The printed translucent screen facing the autobahn is indirectly lit by halogen lamps reflected from a white reflection membrane mounted on the wall of the hall. To avoid glare for drivers, the maximum allowable illuminance is 100 lux. In order to create visual links and a sense of spatial generosity, the intermediate glazing was designed to be as transparent as possible. In order to create a facade as homogenous as possible and to reduce costs, we selected conventional facade elements and avoided the use of cover plates. It was necessary to execute the dividing wall between the offices and the hall in toughened glass as a safety measure for the walkway in the hall. The wall is scarcely visible, being vertically jointed with silicone, and frameless detailing having been carried out at the upper and lower junctions. The glass doors were also fitted with glass door hardware. The printed net to the south facade filters views out to the landscape in a soft-focus manner and the structural tension system is placed out of sight of the building's users, being located above the ceiling height and below the grating level. As a result of the dimensions of the weave and the dark coating on the inside of the net, it is barely visible when viewed in front of a light-coloured background and is only noticeable when viewed from an oblique angle.

Low budget – high performance

Although originally planned as a steel framed structure, the costs involved in providing the necessary 90-minute fire-resistant coating for the storage space caused the building to be redesigned as a more economical reinforced concrete structure. Because the concrete columns are only visible through the elevated window strips and the storage space is clad in opaque panels, the architectural quality of the design could be maintained.
The column axes were set at 6.2 m centres, in order to provide the contracted company with the most economical structural solution. The structural strength of the ribbed metal sheeting could be fully exploited, thereby avoiding the need for secondary beams.

Minimisation of expensive glass component

Other savings included the omission of additional metal angles and bracing tubes for window fixings. For the closed facade areas, we also specified the cheapest metal panels on the market. On the shorter sides the panels span a clear vertical height of seven metres while on the southern facade they are mounted horizontally. The cost limits were indeed so tight that we had to use standard coloured panels instead of the pure white ones originally proposed. The number and size of the domed roof lights were also reduced to save costs; instead of the original double row of skylights, placed in each axis, they were reduced to a single dome in alternating axes.

Refinement of standard products

Even economical building materials, however, can be used in architecturally satisfying ways, with minor modifications sometimes leading to quite individual solutions. The omission of cover strips for example, and the special treatment of sheet abutments and corner details helped to create a building skin of distinctive character. During negotiations with the contracting firms we were able to reach an agreement that the detailing solutions would not cause any price increases. This was only possible because the individual solutions were developed and optimised in collaboration with the contractors. We also designed the office furniture. Coated composite wood boards were attached to 1.20-metre-high standard shelving with stainless-steel angles and are designed as tables.

Factory Extension in Murcia

Architects: Clavel Arquitectos, Murcia

Upgrading of industrial area

In order to meet demands for more office and showroom space at their plant, in Murcia, a Spanish manufacturer planned an extension to the existing factory building which was located in an industrial area. This was seen as an opportunity to upgrade the industrial area and a chance for the company to represent itself appropriately.

The architects created an entirely new presence by locating the new extension across the front of the existing factory building along the street frontage. The materials used – steel and glass – were also a reference to the products manufactured by the company, who in fact supplied most of the building materials required.

Connection to the existing construction on the ground floor

The new building is two storeys high and partially blends into the ground level of the existing construction. The new showroom in the ground floor merges with the old production hall and storage area. In order to create a large space, uninterrupted by columns, the architects spanned the hall with a 13-metre-long truss. The rooms for the administration are accommodated in the upper level. Access is provided via a staircase at the entrance and an additional single-flight stair leads directly from the upper level into the older building.

Interplay of transparency and translucency

The distinctive feature of the new annexe is its translucent facade of profiled glass. The upper level facade is set back from the front in an irregular, zigzag line while the ground level glass facade is set behind a skin of perforated stainless steel. This envelope not only provides sun shading for the spaces concealed behind, but also bears the company name. An interesting play of light is created by the layering of these different materials which, depending upon the angle of incoming light, shifts between transparency and translucency. During the day the lettering dominates, while at night – enhanced by artificial lighting - the lettering recedes and allows the inner life of the building to be revealed.

roof plan
scale 1:1000
sections • floor plans
scale 1:250

1 extension
2 factory building (existing)
3 showroom and sales
4 conference room
5 entrance
6 office
7 meeting room
8 reception

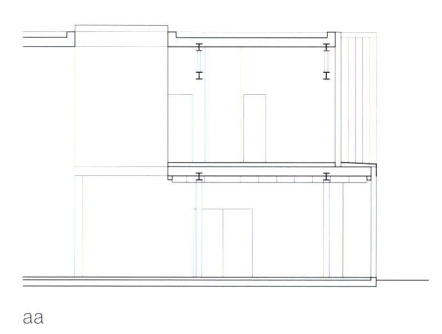

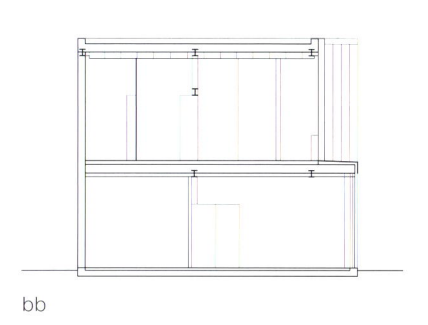

Project details:
Usage:	factory and showroom
Construction:	steel
Total construction costs:	500,000 Euros
Cost per m² usable area:	318 Euros
Total internal volume:	12,000 m³
Total built area:	1,650 m²
Total usable area:	1,572 m²
Built site area:	1,370 m²
Total site area:	6,580 m²
Period of construction:	12 months
Date of construction:	2005

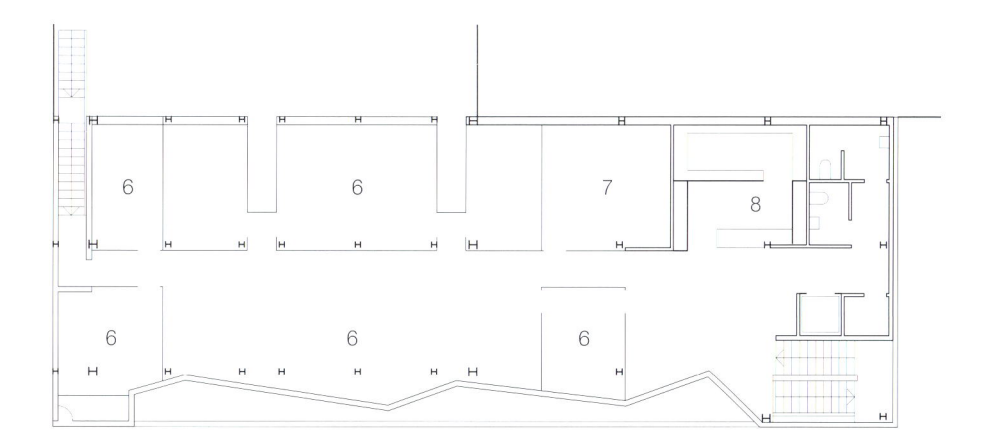

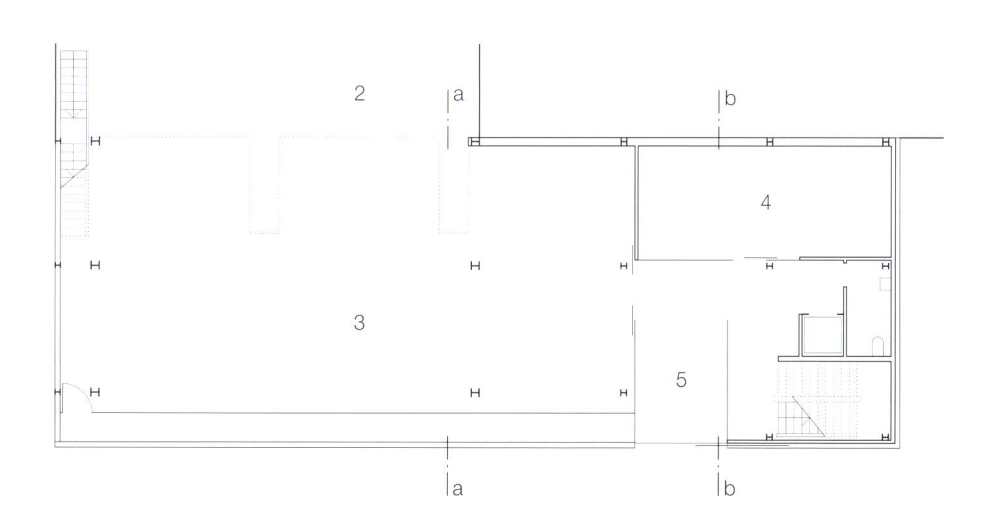

aa   bb

153

vertical section scale 1:20

1. roof construction:
   100 mm gravel
   separation membrane
   50 mm thermal insulation
   separation membrane
   roofing sheet
   max. 250 mm graded concrete
   300 mm composite floor, trapezoidal sheet with concrete topping
2. 140 × 80 mm galvanised aluminium channel section
3. 6 mm channel profile glass, double layer
4. 50 mm steel angle
5. 300 × 100 mm steel channel section
6. 240 mm I-beam,
7. acoustic floor construction:
   12.5 mm plasterboard
   100 mm mineral-wool insulation
8. double glazing: 2× 4 mm laminated safety glass + 10 mm cavity + 2× 4 mm laminated safety glass in galvanised aluminium frame
9. perforated stainless-steel sheet
10. rendered brickwork
11. linoleum
    300 mm composite floor, trapezoidal sheet with concrete topping

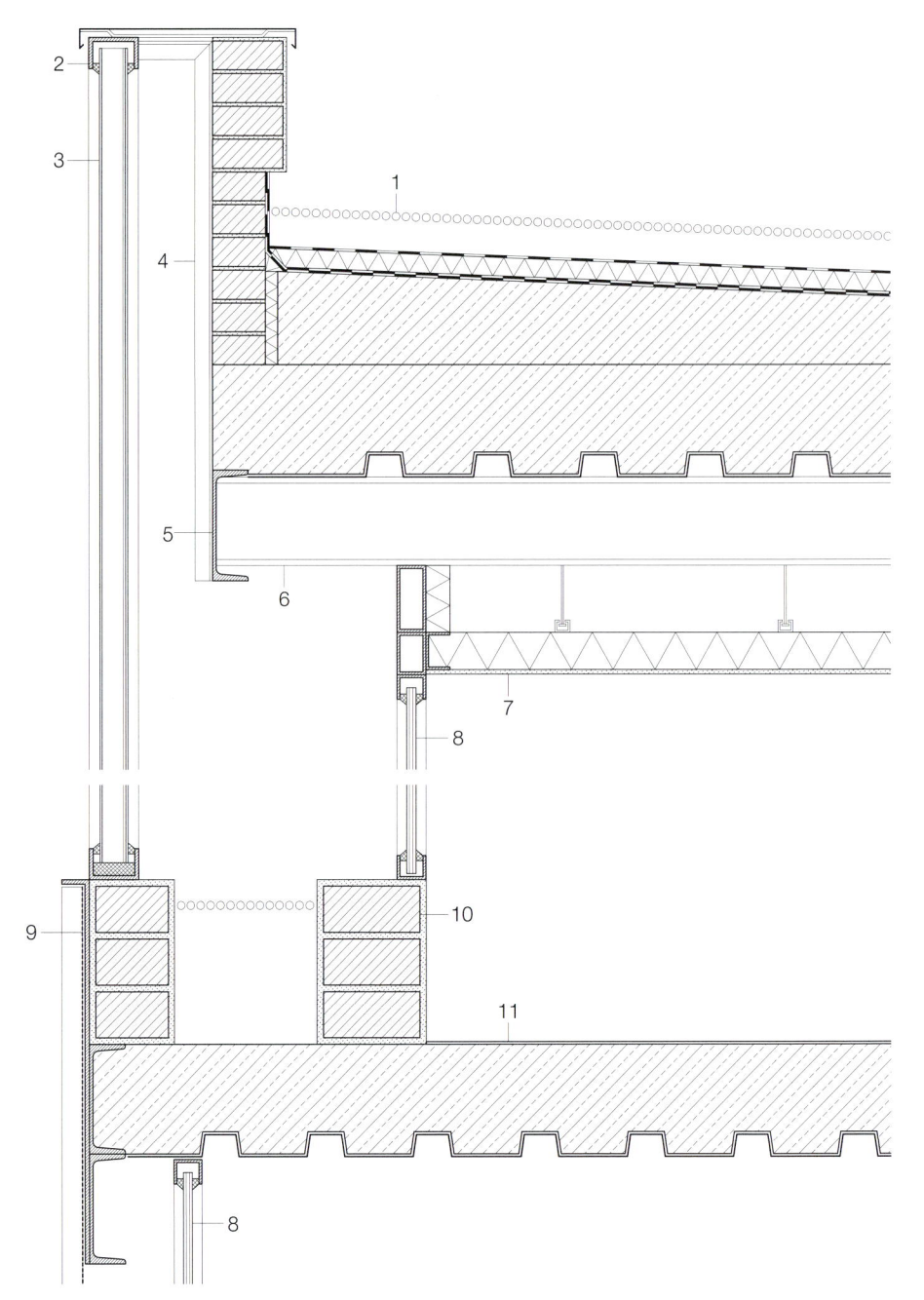

154

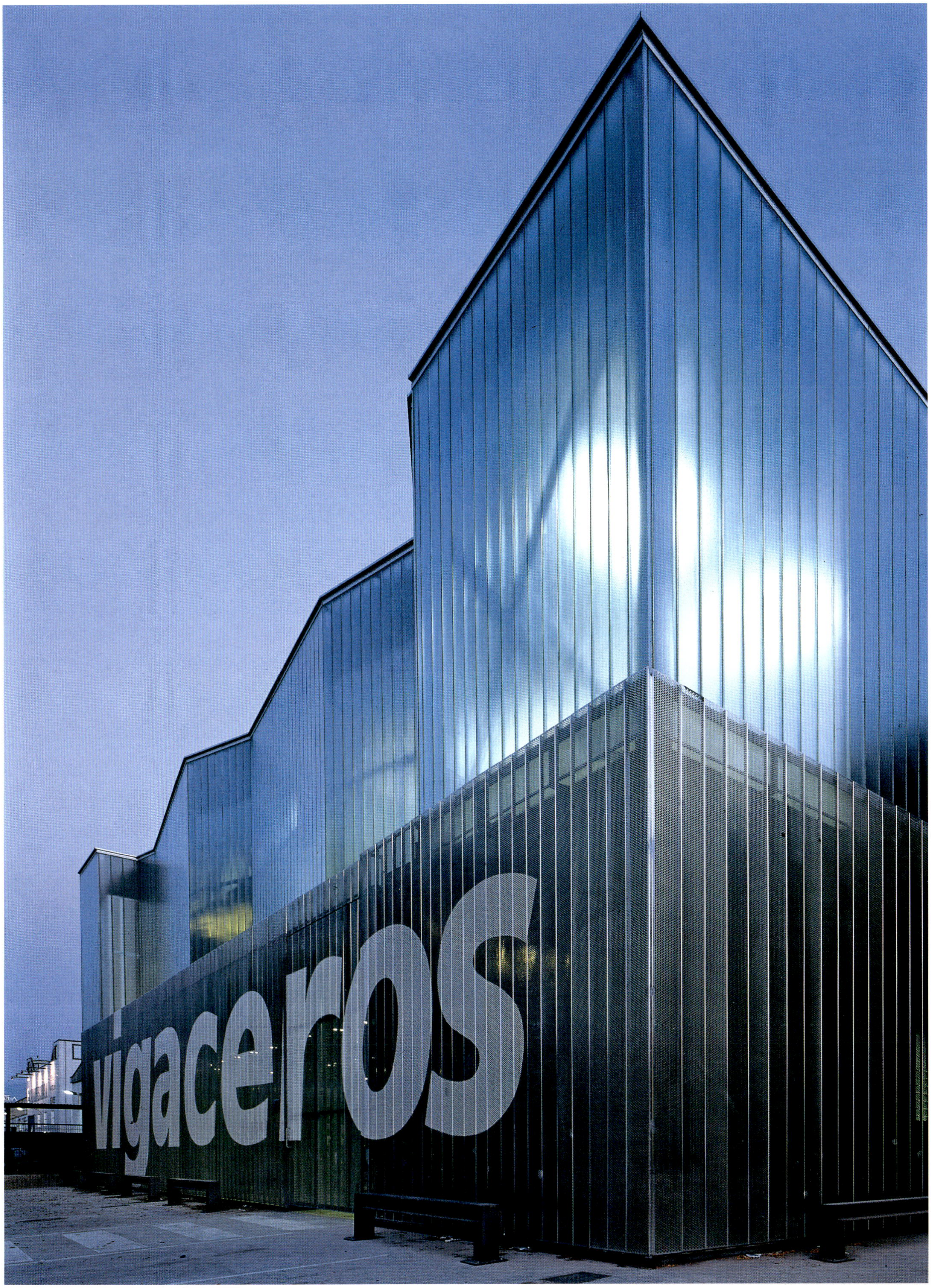
vigaceros

Wiper Factory in Bietigheim-Bissingen

Architects: Ackermann and Partner Architekten, Munich

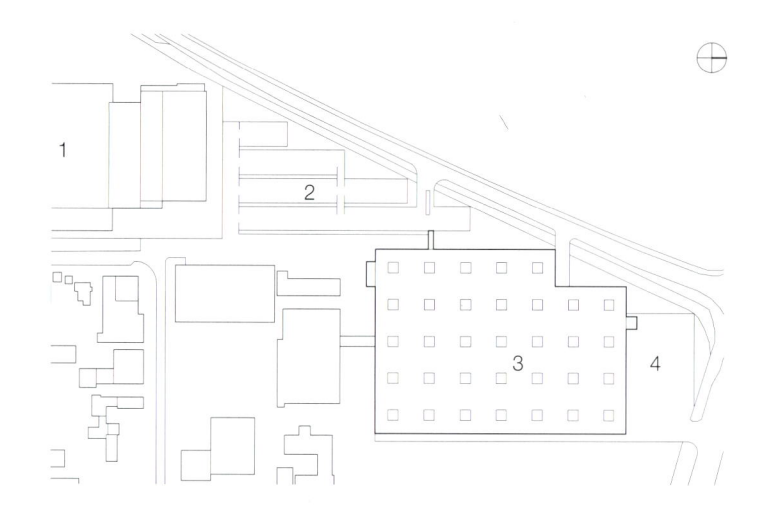

**Modular construction system
Economical serial structure
Restrained appearance**

This company located in the Stuttgart region supplies leading international car manufacturers with windscreen wipers and their operating mechanisms. Directly affected by the economic constraints of the automotive industry, the company is also forced to pay close attention to the cost-efficiency of their built constructions.

A long-term restructuring programme for one of the works and the erection of two new production halls were the subject of a special study. The architects were awarded the scheme for their translation of the company philosophy and for the flexibility of the modular structures they proposed. The restructuring of the production works could be carried out in a number of stages. The fundamental concept of the project was to secure optimal production processes and unimpaired material flows rather than to create an exalted gesture to architectural design. All production units are laid out on a single level; material delivery, manufacturing and dispatch of the finished product. The equal status of production staff and those in administration and development is expressed in a common overall structure and a standardised facade design, with only minor variations reflecting the different functions and representing the logical internal production sequences. Continuous glazing to the facades and partitions, together with the roof lights, ensures a bright, well lit internal environment and simplifies the overall clarity of the scheme. Even special processes requiring air-tight conditions, like the bromide bath tract, can be viewed from the administration department. Light wells extending over three storeys articulate the open plan offices, illuminate the internal zones and create access routes and spatially attractive communication spaces. Flexibility is secured in both the open plan offices and the production hall by the mechanical service runs being located immediately beneath the soffit, where all connections for machinery and appliances are located.

site plan
scale 1:5000

floor plan · sections
scale 1:1000

1 motor works
2 parking area
3 wiper works
4 logistics yard
5 visitors' entrance
6 staff entrance in basement
7 entrance hall
8 administration offices
9 development offices
10 decentralized office units
11 roof lights
12 principal beams
13 link to existing building, goods entrance
14 glazed staff rooms
15 dispatch
16 central control systems
17 extrusion (wiper rubber)
18 bromide bath (wiper rubber)
19 space for planned paint shop
20 deliveries for laboratory and services in basement

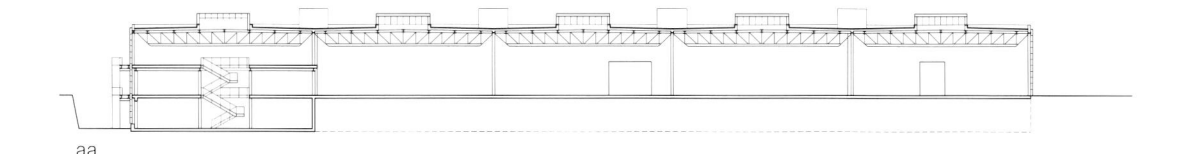

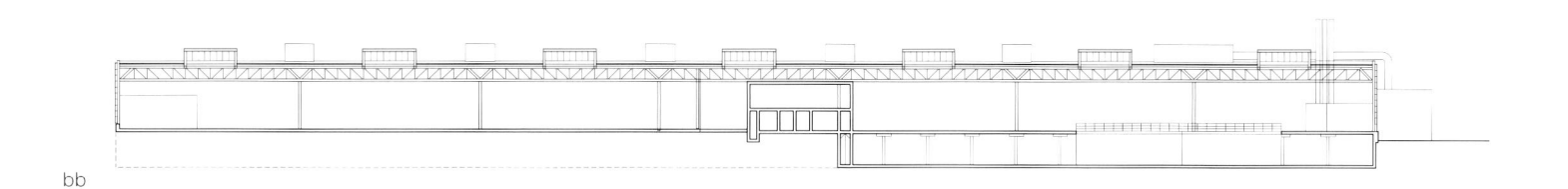

aa

bb

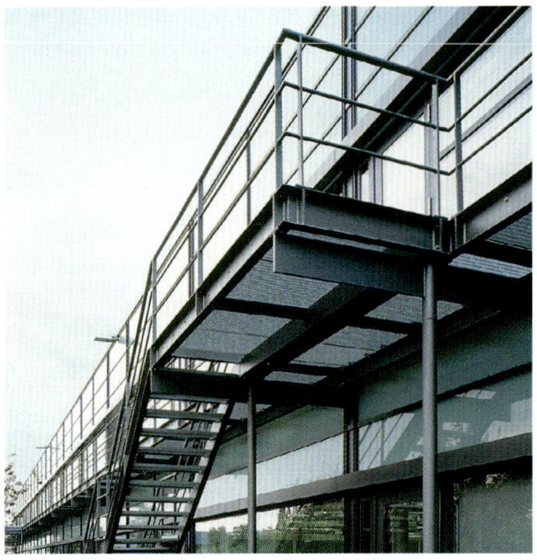

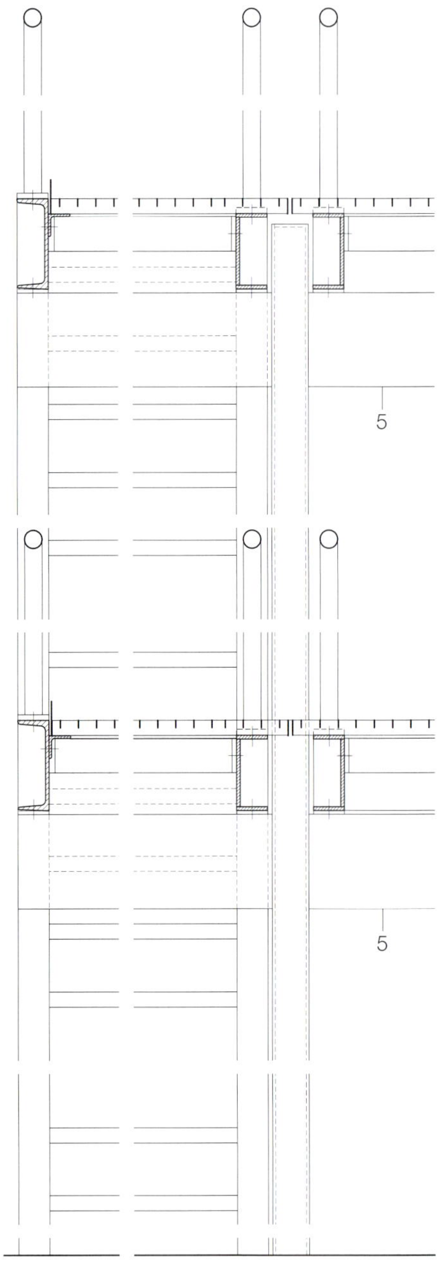

5
5

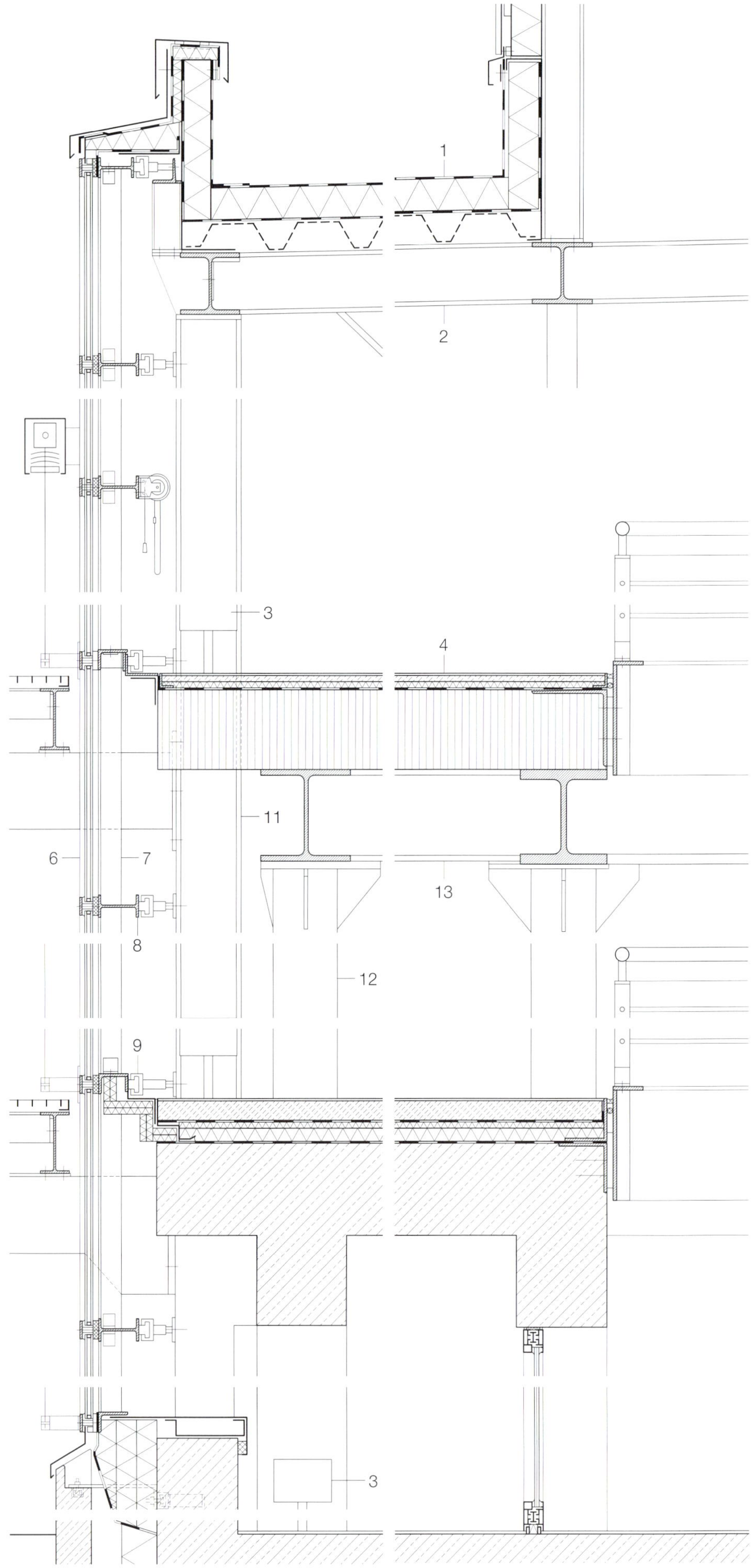

1
2
3
4
11
6
7
13
8
12
9
3

facade section   scale 1:20
details   scale 1:5
1   fluoropolyolefin plastic sealing layer
    120 mm mineral wool, vapour barrier
    1 mm perf. acoustic trapezoidal sheeting
2   200 mm I-beams: upper chord of girder
3   convector
4   5 mm carpeting, 20 mm dry-screed
    20 mm impact-sound insulation, separation
    layer, 260 mm vertically stacked-plank floor
5   250 × 25 mm steel bracket
6   cover strip to aluminium facade
7   70 mm steel T-section post, stressed
8   140 mm steel I-section rail, stressed
9   aluminium section allowing movement
10 stainless-steel threaded bolt
11 primary structure: 220 mm steel I-column
12 secondary structure: Ø 216 × 30 mm steel tube
13 sec. structure: 300 mm steel I-beam

Project details:
Usage: Production, administration
Floor areas: 17,500 m² (office)
                            11,395 m² (admin.)
                            32,470 m² (total)
Internal ceiling height: 4.0/4.5 m (office)
                            9.0 m (admin.)
Maximum height: 13.8 m
External dimensions: 172.7 × 123.5 m
Total internal volume: 237,975 m³
Construction cost: 27.5 million Euros
Cost per m²: 847 Euros
Cost per m³: 115.5 Euros
Live loads: 15/3 kN/m²
Structural span: 24.5 m
Constructional grid: 24.5 × 24.5 m
No. of employees: 600 (production)
Construction period: May 2001–March 2003

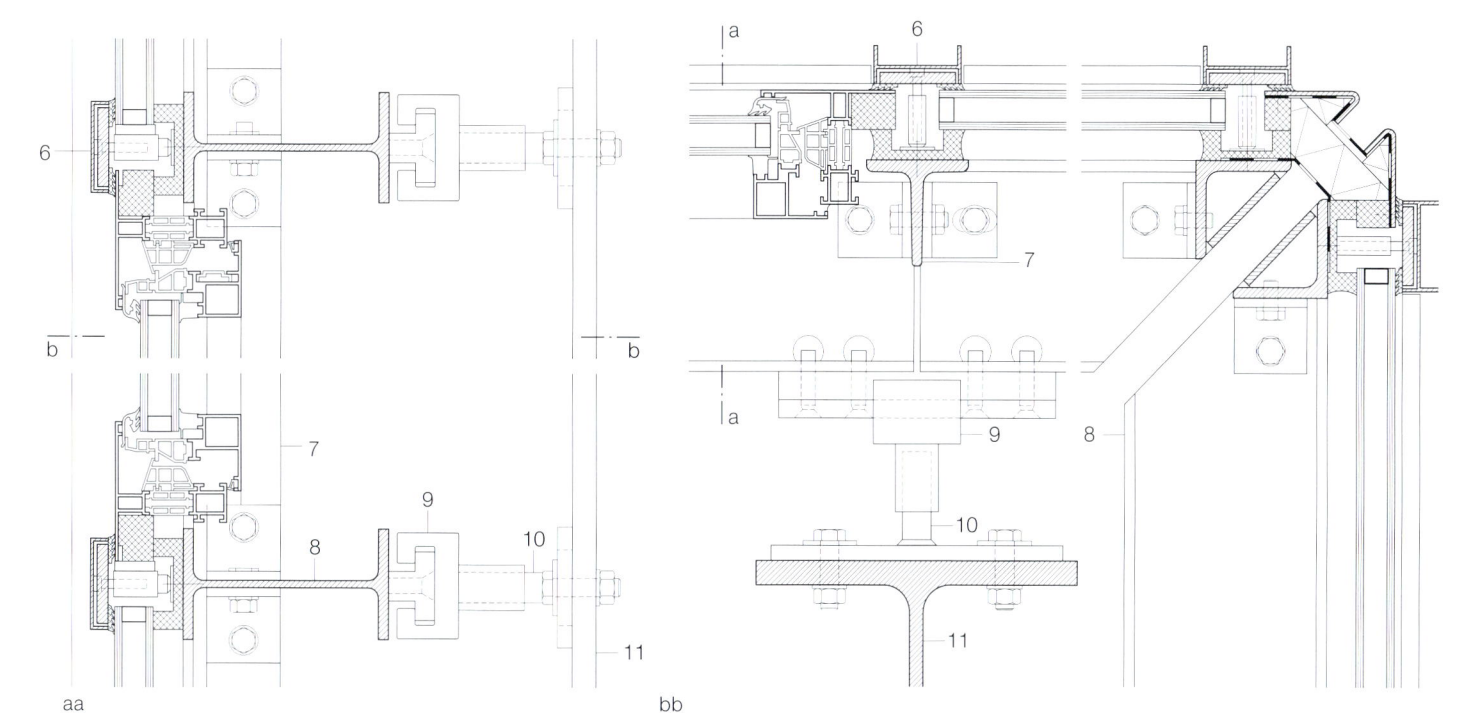

aa   bb

From Urban Planning to Construction Grid – Flexibility as a Design Strategy

Peter Ackermann

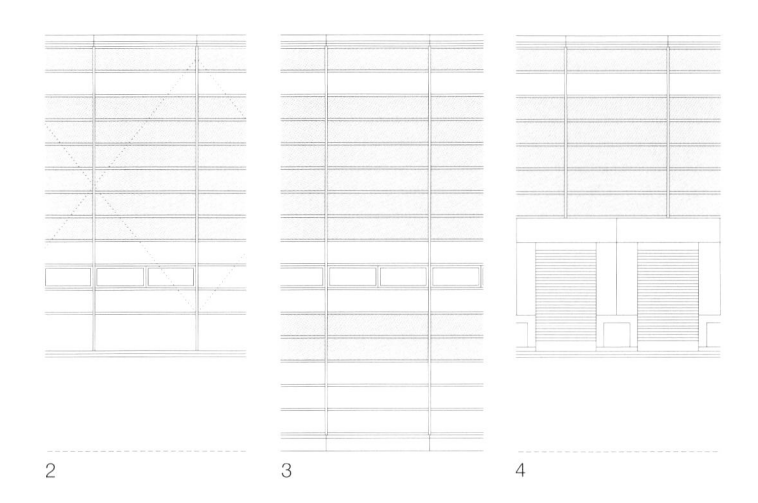

2 3 4

Overall concept rather than formal gesture

Initially, two architectural practices; Clause Vasconi and Ackermann and Partners, were invited by the company's Paris head office to make proposals for the new complex, in collaboration with two planning offices specialising in industrial plants. The selection process was more in the nature of an interview than a competition; the architects were questioned on methods of operation, performance and quality standards. The conceptual formulation was defined after numerous meetings and briefings. Our initial proposal involved schematic concepts relating to the long-term, gradual restructuring of the entire works rather than detailed architectural plans (ill.1).

Flexible modules

The initial task was to plan two discrete plants – one for wipers and one for their motors – on a triangular site. The two independent companies had dissimilar requirements due to their different production sequences and logistics. Neighbouring developments also had to be integrated into the urban planning. The wiper production works is a free-standing volume, while the production plant for the motors is directly connected to an existing five-storey building.

We initially developed a modular structure that allowed a flexible response to existing constructions and the complicated site layout. The 24.5 × 24.5 m grid reflects a number of constraints, including the construction sequence in relation to the size of the site. Functional relationships were also important in terms of the relocation strategy and the logistics of maintaining operations during the construction work. The predominant consideration was the economy of the span dimensions, in relation to flexibility, for the production processes.

Open structures

It ultimately proved possible to implement the spatial concept of a flexible "open working space" in an economical, ecologically friendly building (p. 162 ill. 1). The first construction stage was the production hall, which was in operation for a year before the administration and development modules were added. Within the office construction, a secondary table-like steel structure was inserted, which divides the overall volume into two storeys. All electrics, media and services are, without exception, suspended from the ceilings and constructed as open, visible installations. This, of course, provides maximum flexibility in both the production area and for the office workstations. Electrical and computer services are provided via mobile, suspended cable ducts – similar to spinal columns – which connect the writing desks with the ceilings. Expensive, complicated double or hollow service voids could thus be avoided; a considerable contribution to the economy of the project. The basement level is to contain various laboratories, which are gradually being installed. The paint shop is at present located in an existing building. A boxed-out opening in the floor of the new hall is dimensioned to accommodate any possible plant, regardless of future technology.

Typification down to the Smallest Details – Facade Planning

The continuity of the architectural form called for a facade concept that could be adapted to meet all situations. A fully glazed facade was designed to create an impression of transparency and spaciousness in spite of the dimensions of the scheme. Detailing strategies were necessary in order to

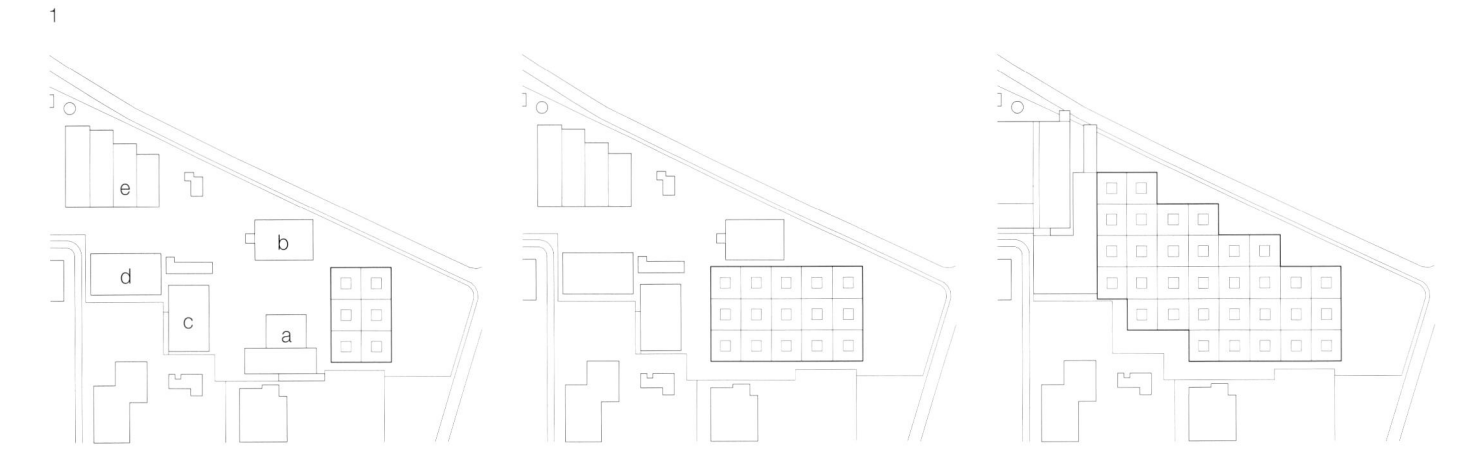

1

1 Feasibility study for selection procedure:
diagrams of step-by-step conversion of works site and
successive incorporation of working areas in the new building
  a existing building (unrelated company)
  b existing wiper production
  c paint shop
  d dispatch
  e existing motor works
2 North facade of production (pale-grey, matt panes), the height of the concrete plinths varies according to ground level.
3 West facade of production with glazed laboratory in basement.
4 Noth facade with loading ramp.

achieve a low-cost solution for such a large proportion of glazing, a cohesive image and simultaneously to communicate the equality of the employees, whether in production, development or administration. The workplaces further away from the facades receive daylight from roof lanterns laid out on a 7 × 7 m grid. In the production hall, the roof lights are vertically glazed at the sides with translucent glass to avoid glare. The integrated smoke and heat ventilation systems are utilised to provide fresh air during mild seasons. The sloped roof lights over the administration areas are glazed with clear glass in order to ensure that sufficient daylight enters the 24.5 metre-deep and 100 metre-long two to three storey space. The ceilings below are cut out to create light wells; stairs have been incorporated into three of the five light wells. These additional access routes enhance communication between employees while providing short interaction paths. Visitors also benefit from visual connection with the upper level and, through the glass partitions, to the production hall.

Modular vertical grid
The undulating topography of the site formed the starting point for the vertical grid. First, the differing levels for the various functions were determined; the main level of the hall, the production floor, is at ±0.00. Access for deliveries is at -1.20 m, and the day lit basement with laboratories, testing facilities, staff rooms and personnel entrance is at -4.50 m. The intermediate office level is at +4.00 m, and the eaves at +8.85 m. By dividing the elevational height of the hall facade into eleven equal elements a versatile unit dimension of 80 cm was attained.

Individual requirements – identical details
This fundamental facade structure based on standard elements could be applied to various situations, irrelevant of differing functions and orientations. In the production hall facade, transparent glass was used up to eye level and beneath the eaves strip. Between these bands of glazing, translucent panes were used to protect against glare (ill. 2). The production hall, which has a basement storey and is oriented to the road, is articulated with transparent glazed strips to coincide with the office facade (ill. 3). In order to fix the steel escape balconies, which are required for maintenance and sun-shading, to the western facade it was deemed necessary to execute the vertical facade elements as double posts. The glazing is obscured to coincide with the eye-level of seated workers. External sun blinds and an internal blind against glare take account of the lighting conditions required for computer work (ill. 4). Although the opening facade elements, which are provided for the intake of air to support smoke extraction and exhaust systems, are not strictly necessary for ventilation purposes, they nevertheless ensure a sense of well-being among employees. At the dispatch area the glass facade is finished at ground level by a precast concrete element which also provides additional impact protection (ill. 2, 4).

Construction
The facade construction consists of rolled steel posts and rails, which were mechanically stressed in order to comply with the exacting tolerances necessary for facade construction. The application of open sections lends it a filigree quality and a sense of elegance. The external sections, which are necessary to ensure thermal separation of the construction, form a standard system and are identical in the office and production areas. Various types of glazing were used to meet different requirements in respect of building physics. A great deal of effort went into designing the interface between the facade construction and the structural steel columns which bear the horizontal loads. A horizontally sliding, vertically adjustable, load-transmitting connection was developed to accommodate tolerances of up to 4 cm in length.

Internal facades and inserted glass cubes
The internal partition grid reflects the external facade dimensions. Fire protection necessitated the separation of the office and production areas. The installed glazed partition with wired plate glass provides a fire resistance of 30 minutes and also acts as an acoustic screen with a rating of 42 dB – a highly economical solution. It was essential to reduce the horizontal grid dimension of 3.5 m to 1.75 m as fire-rated glazing was not available in these widths. Due to stringent ventilation and security requirements it was deemed necessary to separate the extrusion area from that of the production with single layer glazing. Identical glass partitions were applied for the subdivision of specific rooms; meeting rooms in the offices, decentralised offices and staff rooms in the production area, service rooms and the bromide bath zone within the extrusion area. This enhances the cohesive appearance of production and administration. The standardisation of elements enabled large production runs and an economical form of prefabrication to be practicable.

The Optimisation of the Load-Bearing Structure
Christoph Ackermann

1 First building stage: open work space
  production hall prior to the expansion by the administration areas
2 Assembly of the main beams as continuous girders
  in the longitudinal direction of the building
3 Different loads resulting from roof lights and service runs
  require steel structural members of different cross-sections
4 Isometric view of structure as executed:
  ventilation ducts pass through the main beams
  via V-shaped openings over the columns

From a non-directional to a directional system
The load-bearing structure was initially conceived as a non-directional beam system laid out to a 24.5 metre square grid which could be extended in any direction and would provide adequate column-free space for the production. The roof structure comprised a system of primary and secondary beams and each of the roof modules was to be borne by four hinged columns.

In the course of the planning, however, the media lines for services and production technology developed into dense clusters, so that the loading was concentrated along certain routes and was no longer evenly distributed over the area of the structure. A weight of up to 350 kg/m accrued. The load-bearing behaviour of non-directional structural systems is no longer necessarily biaxial when subject to linear loading. In other words, the structure would have been over-dimensioned in one direction and no longer strictly economical. The non-directional bearing principle was abandoned, therefore, although the 24.5 metre square column grid was retained (ill. 4).

Coordination with technical services
To optimise the load-bearing structure, the layouts for the mechanical services and the production technology were superimposed on the structural plans. The loads on individual areas were calculated, and the spacings of the secondary beams adjusted accordingly. A continuous structural system of primary beams running the length of the production hall, with suspended, single-bay secondary beams, was developed in a step-by-step planning process. The individual elements were consistently adapted to the loads and their dimensions altered. Elements subject solely to tension stresses were fabricated from steel plates; for those subject to tension and

2

compression, narrow channel sections were used; while compression members liable to buckling were formed from rolled steel sections symmetrical about two axes. Thus it was possible to produce well-proportioned, high performance elements with optimal load-bearing behaviour. This resulted in material savings and also made the function of the various elements visible. The structural load-bearing collection points are simultaneously the service distribution points.

Bracing
The roof plane was designed as a plate structure with prestressed steel rod bracing. It serves to stabilise the upper chords of the lattice girders and to transmit wind loads to the concrete cores and vertical bracing. The lower chords of the continuous beams are further braced against buckling near the columns. In the longitudinal direction, the bracing ele-

3

ments are located in the facade plane in the middle of the 171.50 metre-long hall. These elements allow the load-bearing structure to expand outwards in both directions from the centre, thereby effectively halving the extension length. As result, it was possible to construct the hall without expansion joints and additional diagonal bracing in the production areas. A continuous girder system, with a maximum span of 98 metres, provides bracing in the longitudinal direction.

The administration tract is separated from the production hall by a construction joint and is independently braced by stiffening elements and by fixing the roof plate to the concrete cores, in order to avoid any temperature related stresses between these two areas. The inserted steel table-structure – with a stacked-plank floor acting as an intermediate level in the offices – is braced by the framing effect on both sides and flexibly connected to the columns that support the roof structure.

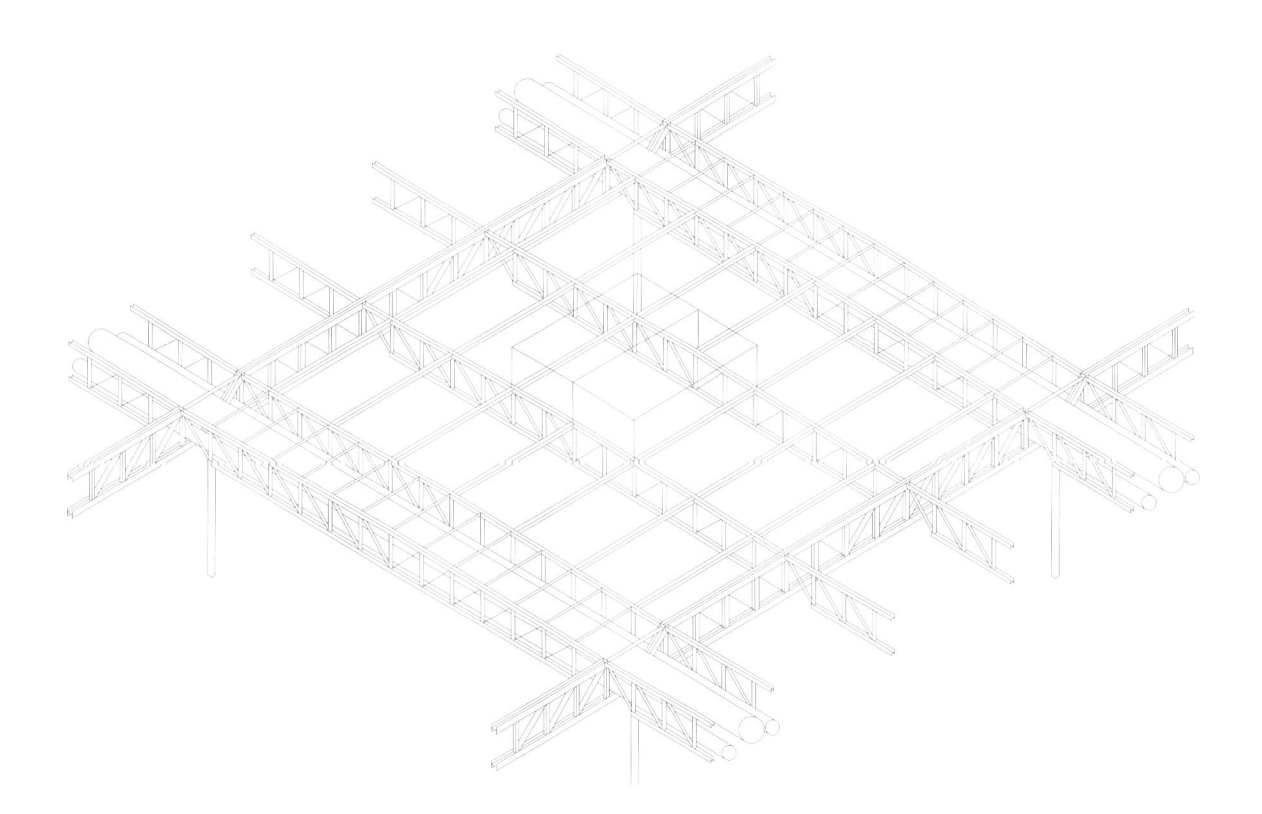

4

The Building in Operation
Helmut Bucher

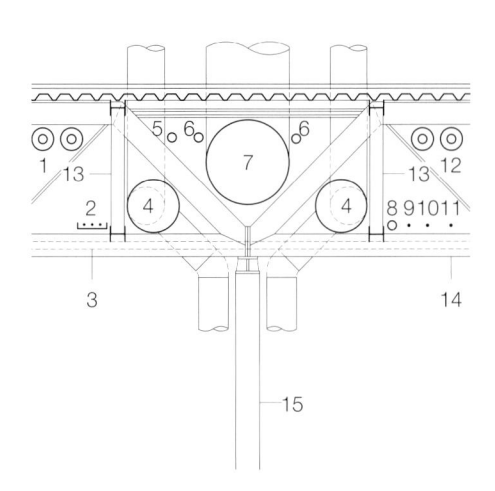

Clearly arranged service installations
In order to arrange the building's technical services clearly the clients requested that the service runs be installed in a single layer, intersection points be clearly detailed and future replacement of plant and appliances to be possible without major changes in structure or other installations. Uniform appearance and maximum flexibility in the offices and the hall was ensured by distributing all mechanical services from the soffit (ill. 1). The air-conditioned office spaces required a higher density of cables, decentralised electrical and data-distribution stations and different light fittings than the production hall. Decentralised ventilation equipment on the roof allows for flexible control of the system and provides extra capacity. Exhaust air from machinery is also removed by decentralised ventilation units and expelled above the roof. All pollutants are dealt with by appropriate systems e.g. thermal post-combustion units.

Technical services centre
All Valeo works throughout the world are subject to the same safety standards. The sprinkler installation allowed larger fire compartments, which in turn facilitated flexibility and openness. The on-site water supply for fighting fires is sufficient for the entire site, not just the building itself. All supply and security systems are located in the basement in a discrete, secure department. The mains power, transformers and medium voltage supply are also found in the basement and have additional external access.

Building control centre
The building control centre is situated at the south-eastern corner of the hall. All electrical and media flows are monitored and distributed via an energy-cost management system. Disturbances of security relevant systems like sprinklers, smoke alarms, door monitors, emergency power supply and lifts are registered by an emergency control system and information is immediately forwarded to the stand-by maintenance system. An integrated peak-demand optimiser enables the use of an electricity power limiter.

Plant dimensions
Approximately 465,000 m³/h of partly cooled air can be supplied to the production areas and offices by 15 decentralised ventilation installations and roughly 6 km of metal ducts. All ventilation plant is coupled with heat-recovery systems. Two 1,065-kW cooling machines and two cooling towers supply the works with cold water. The production process is up to 60 per cent dependent on the use of compressed air, which is provided by three compressors. The heating supply is based on two gas-fired hot-water generators with a total capacity of roughly 2,000 kW. An emergency generator and four 1,200-kW transformers ensure the power supply for the works.

Employee acceptance
This spacious, brightly lit structure has found broad approval among the staff. Even the open-plan offices have now been accepted, as they promote a communicative working atmosphere. The glazed areas ensure a high degree of transparency, but can also result in glare, so that additional screening is planned. Acoustic problems have been resolved by aesthetically pleasing methods; a timber soffit has been incorporated above the ground floor and perforated ribbed metal sheeting fixed to the ceiling over the upper floor. The glazed recreational rooms in the production hall have also proven to be very popular.

Section through service runs
scale 1:100

1 cooling Ø 80 mm (nom.)
2 electrical runs
3 compressed air Ø 100 mm (nom.)
4 air supply Ø 700 mm
5 rainwater pipe
6 sprinkler runs
7 air extract Ø 1120 mm
8 waste water Ø 100 mm
9 circulation Ø 15 mm
10 drinking water (cold) Ø 20 mm (nom.)
11 drinking water (warm) Ø 20 mm (nom.)
12 heating Ø 65 mm (nom.)
13 secondary beam
14 main girder
15 column

Helmut Bucher is the head of facility management for Valeo Wiper Systems, GmbH, Bietigheim

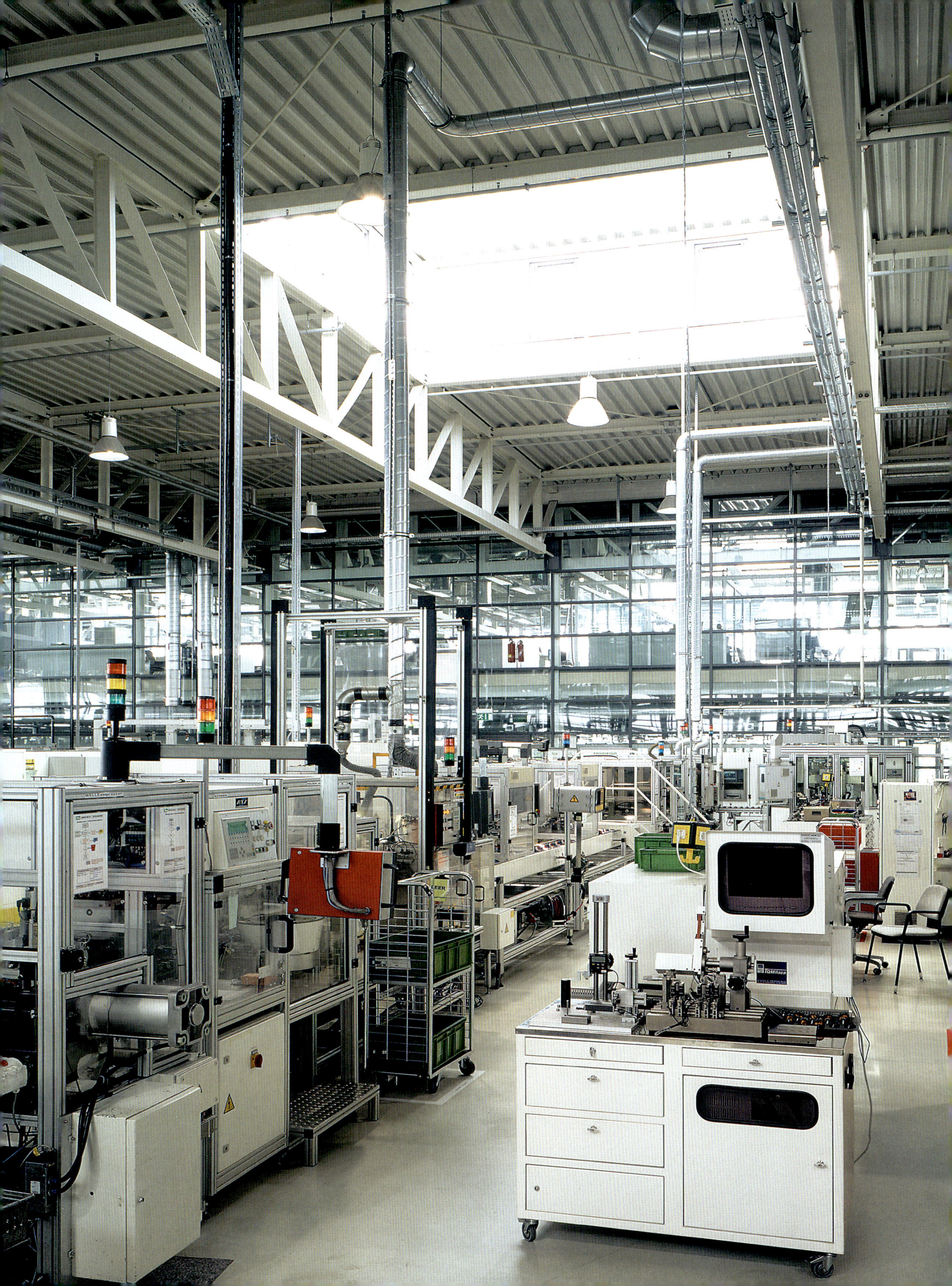

Architects – Project details

Showroom near Zevenbergen

Client: Roma Rollladen + Tore GmbH, Burgau
Architects: ott architekten, Augsburg
Wolfgang Ott
Project leader: Dietrich Bürgener
With: Annabelle Schmid, Christoph Katzenberger
Project management: Rainer Eichelbrönner, Poppenhausen
Structural engineer: Ingenieurbüro Schütz, Kempten
Gerhard Pahl, Kempten
Date of construction: 2006

info@ott-arch.de
www.ott-arch.de

Wolfgang Ott
Born 1961 in Augsburg; studied architecture at the University of Karlsruhe and at the University of Stuttgart;
1990–1994 employed by Behnisch & Partner, Stuttgart; 1995 employed by Peter Hübner, Neckartenzlingen;

since 1997 Ott Architekten, Augsburg in partnership with Ulrike Seeger

Studio in Madrid

Clients: Luis Gordillo & Pilar Linares, Madrid
Architects: Ábalos & Herreros, Madrid
Structural engineer: Juan Gómez, Madrid
Landscapeplanner: Fernando Valero, Madrid
Date of construction: 1999

studio@abalos-herreros.com
www.abalos-herreros.com

Iñaki Abalos Vázquez
Born 1956 in San Sebastián, Spain; 1978 University degree from the Escuela téchnica superior de arquitectura de Madrid (ETSAM); 1984–1988 lecturer at ETSAM; 1991 Ph.D. from ETSAM; 1992 professor at the faculty Proyectos Arquitectónicos at ETSAM

Juan Herreros Guerra
Born 1958 in San Lorenzo de El Escorial, Spain; 1984–1988 lecturer at ETSAM; 1985 University degree from ETSAM; 1994 Ph.D. from ETSAM; 1995 professor at the faculty Proyectos Arquitectónicos at ETSAM

1985 establishment of the practice Ábalos & Herreros

Pharmacy and Medical Practice in Plancher-Bas

Client: SCI du Rahin, D. Lachat
Architects: Rachel Amiot & Vincent Lombard, Besançon
Structural engineer:
Jean Luc Sandoz, CBS-CBT, Lausanne (timber construction)
F. Durant, F.D.I., (concrete construction)
Date of construction: 2004

architectures-amiot-lombard@wandoo.fr

Rachel Amiot
Born 1965 in Vesoul, France; 1990 degree from the Ecole d'Architecture de Nancy

Vincent Lombard
Born 1964 in Lure, France; 1990 degree from the Ecole d'Architecture de Nancy

www.cbs-cbt.com

Jean-Luc Sandoz
1993–1998 professor for timber construction at the Ecole Polytechnique Fédérale de Lausanne; teaching positions at various European universities;
since 1998 owner of the engineering practice CBS-CBT (Concept Bois Structure – Concept Bois Technologie) in Switzerland and France.

Mobile House in England

Client: Tim Pyne, London
Architects: Mae Architects, London
Project leader: Michael Howe
Structural engineer:
Techniker, London
Atelier one, London
Building contractor:
Discovery Contractors Ltd., Kent
Date of construction: 2002

office@mae-llp.co.uk
www.mae-llp.co.uk
www.timpyne.com

Alex Ely
Employed by Pierre d'Avoine Architects; co-owner of Mae Architects LLP, London

Michael Howe
Studied at the University of Westminster; employed by Patel Taylor Architects, Zaha Hadid Architects and Matthew Lloyd Architects; teaching position at the University of Greenwich;
co-owner of Mae Architects LLP, London

Prefabricated House from Denmark

Architects: ONV arkitekter, Vanløse
With: Søren Rasmussen,
Christian Hanak
Project management:
Hp gruppen A/S, Hjørring
Structural engineer:
DPR ingeniørerne a/s, Hjørring
Jens Abildgård

onv@onv.dk
www.onv.dk

Søren Rasmussen
Born 1956; 1982 Bachelor of Architecture, Technology and Construction Management from the Blekinge Tekniska Högskola in Karlskrona, Sweden;
1988 Master of Applied Arts;
1986–2000 employed in various architectural offices;

since 2000 owner of ONV arkitekter, Vanløse

Straw House in Eschenz

Client: Stokholm Family, Eschenz
Architect: Felix Jerusalem, Zurich
Project management:
Turi Weiss, Stein am Rhein
Structural engineers:
SJB Kempter Fitze AG, Frauenfeld
and Crétion Holz, Herisau
Date of construction: 2005

f.jerusalem@bluewin.de

Felix Jerusalem
Born 1963 in Freiburg im Breisgau;
1991 degree from the ETH Zurich;
1991–2000 employed by various offices in New York and Zurich;

since 1992 own office in Zurich

House in Shimane

Client: private
Architects: Sambuichi Architects, Hiroshima
Project leader: Hiroshi Sambuichi
With: Hindenori Ejima, Manabu Aritsuka
Structural engineers:
Arup Japan, Tokyo and
Rinken Corporation, Shimane
Date od construction: 2005

samb@d2.dion.ne.jp

Hiroshi Sambuichi
Born 1968 in Japan; degree 1992 from the Technical University in Tokyo; 1992–1996 employed by Ogawa Shinichi Atelier in Hiroshima; teaching position at the Yamaguchi University, Yoshida;

since 1997 Sambuichi Architects, Hiroshima

House in Aitrach

Clients: Marion and Ralf Hinterberger, Aitrach
Architects: SoHo Architektur, Augsburg
Project leader:
Michael Weizenegger
With: Philip Sohn
Structural engineering and building physics: Ingenieurbüro Herz & Lang, Weitnau
Date of construction: 2006

info@soho-architektur.de
www.soho-architektur.de

Alexander Nägele
Born 1970 in Memmingen;
1999 degree from the Institute of Technology Augsburg;

2000 establishment of SoHo Architektur together with Jörg Schiessler in Augsburg;
2007 transfer of the practice to Memmingen

Weekend House in St Andrews Beach

Client: D. McNair
Architects:
Sean Godsell Architects, Melbourne
With: Hayley Franklin
Structural engineer:
Felicetti, Melbourne
Date of construction: 2006

godsell@netspace.net.au
www.seangodsell.com

Sean Godsell
Born 1960 in Melbourne; 1984 degree from the Royal Melbourne Institue of Technology, Melbourne; 1986–1988 employed by Sir Denys Lasdun, London; 1989 employed by The Hassell Group, Melbourne; 1999 Master of Architecture from RMIT, Melbourne; teaching positions in the USA and London;

1994 establishment of Godsell Associates Pty Ltd Architects;

House on Lake Laka

Client: private
Architect: Peter Kuczia, Osnabruck
General contractor:
Kombud, Pszczyna
Solar technology:
Semar, Bielsko-Biala
Green roof: Xeroflor, Leszno Gorne
Date of construction: 2007

kucia@web.de

Peter Kuczia
Born in Poland; degree from the Technical University Gliwice, Poland; since 1992 employed by various offices in Germany; since 1999 employed by agn Generalplaner, Ibbenbüren; since 1998 free-lance work in Poland; numerous articles and presentations

Housing Development in Neu-Ulm

Client: NUWOG Wohnungsgesellschaft der Stadt Neu-Ulm GmbH
Architects: G.A.S. Sahner, Stuttgart
With: Torsten Belli, Jürgen Sick, Susanne Werner, Herrmann Falch
Structural engineerr: Ingenieurbüro Müller, Kirchberg/Iller
Technical and electrical services: Ingenieurbüro Spleis, Laupheim
General contractor:
Schweizer OHG, Westerheim
Date of construction: 2000

gas.sahner-sahner@t-online.de
www.gas.sahner-architekten.de

Georg Sahner
Born 1955; 1983 degree from the University of Stuttgart; 1986–1987 teaching position at the Institute of Technology Biberach; since 1986 collaboration with Klara Sahner; since 2001 professor at the Institute of Technology Augsburg; since 2006 head of masters studies in Energy Efficient Design at the Institute of Technology Augsburg

Klara Sahner
Born 1955; 1983 degree from the University of Stuttgart; since 1986 collaboration with Georg Sahner

Terrace Housing in Milton Keynes

Client: English Partnerships with Taylor Wimpey
Architect: Rogers Stirk Harbour + Partners, London
Structural engineer and general contractor: Wood Newton Ltd
Date of construction: 2007

www.rsh-p.com

Richard Rogers
Born 1933 in Florence;
1954–1959 studied at Yale University; 2004 professor at the Tongji University, China;
2007 awarded the Pritzker Prize

Ivan Harbour
Born 1962 in Irvine, Scotland; 1985 degree from the Bartlett School of Architecture and Planning, London; 1985–1993 employed by Richard Rogers Partnership, London; since 1993 Senior Director at Rogers Stirk Harbour + Partners, London

Andrew Partridge
Born 1961 in Cardiff,
1985 degree from Leicester Polytechnic; 1989 employed by Richard Rogers Partnership, London; seit 2001 associate at Rogers Stirk Harbour + Partners, London

Apartment House in Dortmund

Clients: Uta und Karl-Heinz Klenke, Dortmund
Architects: ArchiFactory.de, Bochum
With: Carsten Deis,
Kerstin van Treeck
Structural engineer: Assmann – beraten und planen, Dortmund
Date of construction: 2004

www.archifactory.de
office@archifactory.de

Matthias Herrmann,
Born 1966 in Tuttlingen, Germany; 1992 degree from the Institute of Technology in Bochum; 1995 degree from the University of Dortmund; employed by Joseph Paul Kleihues, Berlin

Matthias Koch,
Born 1963; 1985–1987 cabinet maker apprenticeship in Dortmund; 1993 degree from the Academy of Fine Arts in Stuttgart ; 1993–1995 employed by Gerber Architects, Dortmund; 1995 employed by Josef Paul Kleihues, Berlin

1999 establishment of ArchiFactory.de

Housing in London

Client: Peabody Trust, London
Architects: Ash Sakula Architects, London
Project manager:
Sandwood Construction, London
With: Cany Ash, Robert Sakula, Duncan Holmes
Structural engineer:
Whitby Bird Engineers, London
Artist: Vinita Hassard, London
Date of construction: 2004

info@ashsak.com
www.ashsak.com

Cany Ash
Employed by GLC architects, Shah-Alam, Malaysia; employed by Burrell Foley Fischer, London, New York und Berlin; teaching positions at various schools of architecture

Robert Sakula
Studied at the University of Liverpool; employed by Sir Clough Williams-Ellis in Portmeirion and by DEGW Architects in London; teaching position at the University of North London and the University of East London

1994 establishment of Ash Sakula Architects in Clerkenwell

Apartment Building in London

Client: Peabody Trust, London
Architects: Niall McLaughlin Architects, London
With: Gus Lewis, Sandra Coppin, Bev Dockray
Project management:
Walker Management, London
Structural engineering:
Whitby Bird Engineers, London
Artist: Martin Richman, London
Date of construction: 2004

info@niallmclaughlin.com
www.niallmclaughlin.com

Niall McLaughlin
Born 1962 in Geneva, Switzerland; 1979–1987 studied at University College in Dublin; since 1990 Niall McLaughlin Architects; 1990–1996 lecturer at the Oxford Brookes University; since 1994 lecturer at the University College in London

Multi-storey Housing in Munich

Client: GBWAG Bayerische
Wohnungs Aktiengesellschaft,
München
Architects: Hierl Architekten,
Munich
Rudolf Hierl
With: Maurice Maync,
Carolin Semtner, Michael Feil,
Tobias Miazga, Andreas Rackl
Structural engineering:
IB Kaspar & Neumann, Munich
Date of construction: 2005

info@hierlarchitekten.de
www.hierlarchitekten.de

Rudolf Hierl
Born 1958 in Neumarkt/Oberpfalz;
1978–1979 studied Germanistics
und Theater Science in Erlangen;
1979–1984 studied Architecture at
the Technical University Berlin;
1985 degree; 1985–1986 employed
by Otto Steidle, Munich; 1986–1987
employed by Paolo Nestler, Munich;
since 1987 own office in Munich;
1989 Ph.D. from the Philipps-
University Marburg; since 1995
teaching position at the Institute
of Technology Regensburg

Hall of Residence in Amsterdam

Client: Wohnbaugenossenschaft
De Principaal
Architects: Claus en Kaan
Architecten, Amsterdam/Rotterdam
Felix Claus, Jaap Gräber
With: Roland Rens
Structural engineering: Van Rossum
Raadgevende Ingenieurs,
Amsterdam
Date of construction: 2002

cka@cka.nl
www.clausenkaan.com

Felix Claus
Born 1956; professor for Architecture at the ETH Zurich; teaching
position at the ETSAM, Madrid

Kees Kaan
Born 1961

1987 establishment of Claus en
Kaan Architecten; the practice
has offices in Amsterdam and
Rotterdam

Youth Camp in Passail

Client: Marktgemeinde Passail
Architects: Holzbox ZT-GmbH,
Innsbruck, Erich Strolz,
Armin Kathan
With: Ferdinand Reiter, Bernhard
Geiger, Marlene Gesierich, Martin
Grafenauer, Christian Haag, Peter
Zelger
Structural engineering: JR Consult,
Johann Riebenbauer, Graz
(timber construction),
Michael Vatter, Gleisdorf
(concrete construction)
Date of construction: 2004

mailbox@holzbox.at
www.holzbox.at

Erich Strolz
Born 1959 in Warth-Hochkrumm-
bach; studied Architecture at the
TU Graz und University of Innsbruck; employed in architectural
offices in Austria, Germany and the
USA;

Armin Kathan
Born 1961 in Lech; studied architecture at the University of Innsbruck and the Academy of Applied
Arts in Vienna; worked in architectural offices in Austria and Germany
and the USA;

since 1993 combined office in
Innsbruck

Ferdinand Reiter
Born 1969; studied architecture at
the University of Innsbruck; since
1993 with Kathan & Strolz/Holzbox

Hotel in Groningen

Client: Nijestee Vastgoed,
Dhr. Renken, Groningen
Architects: Foreign Office Architects, London; Farshid Moussavi, Alejandro Zaera Polo
With: Shokan Endo, Kazutoshi Imanaga, Kensuke Kishikawa, Yasuhisa Kikushi, Izumi Kobayashi, Kenichi Matsuzawa, Tomofumi Nagayama, Xavier Ortiz, Lluis Viu Rebes, Keisuke Tamura
Structural engineering:
Structure Design Group, London
Arup, London
Date of construction: 2001

mail@f-o-a.net
www.f-o-a.net

Farshid Moussavi
Studied at the University of Cambridge, University College London and University of Dundee; 1988 employed by Renzo Piano Building Workshop, Genova; 1991–93 employed by OMA, Rotterdam; teaching position at various universities in Europe and USA; since 2006 professor at the Harvard University of Cambridge

Alejandro Zaera Polo
Studied at the Harvard University of Cambridge and the ETSAM; 1991–92 employed by OMA, Rotterdam; teaching position at various universities in Europe and USA; since 2002 Dean at the Berlage Institute in Rotterdam

1995 establishment of Foreign Office Architects in London

Cultural Centre in Munich

Client: Bürgerzentrum Trudering e.V., München
Architects: Ingrid Amann Architekten, Munich, with Rainer Gittel
Project leader: Stephan Feldmaier
With: Maria Zach
Project management: ARGE Kulturzentrum Trudering; Gebr. Donhauser Hoch- und Tiefbau Unternehmung GmbH & Co, Schwandorf; Architectural Office Franz Xaver Hafner, Munich; Hemmerlein Ingenieurbau (Structural engineering), Bruck i.d. Oberpfalz, Markus Erhardt
Date of construction: 2005

mail@amannarchitekten.de

Ingrid Amann
Studied archeology at the University of Regensburg; studied architecture at the TU Munich; 1995–1999 employed by various architectural offices in Munich und Berlin; since 1999 free-lance architect in Munich;
1999 teaching position at the Institute of Technology Regensburg;
2000 guest lecturer at ther Techncal University of Hannover; 2003 teaching position at the Institute of Technology Munich;
2005 guest professor at the Institute of Technology Munich

Secondary School in Brixlegg

Client: Immobilien Brixlegg GmbH, Brixlegg
Architect: Raimund Rainer, Innsbruck
Project leader: Manuel Breu
Project management:
Alfred Neuner, St. Johann in Tirol
Structural engineering:
ZSZ Ingenieure, Innsbruck
Date of construction: 2007

office@architekt-rainer.at
www.architekt-rainer.at

Raimund Rainer
Born 1956 in Schwaz; 1976–1984 studied at the Technical University in Innsbruck; 1980–1981 international studies at the Technical Institute in Delft und Technion in Haifa; since 1984 own office in Innsbruck

Secondary School in Eching

Client: Landkreis Freising
Architects: Diezinger & Kramer, Eichstätt
Norbert Diezinger, Gerhard Kramer
Project leader:
Johannes Schulz-Hess
With: Diana Hollacher,
Markus Knaller, Marcel Wendlik
Structural engineering:
Ostermair + Pollich, Freising
Date of construction: 2006

architekten@diezingerkramer.de
www.diezingerkramer.de

Norbert Diezinger
1982 degree from the University of Stuttgart;1982–88 employed by Karl Joseph Schattner, Eichstätt; 1987 establishment of own office in Eichstätt; since 1990 partnership with Gerhard Kramer

Gerhard Kramer
1979 degree from the University of Stuttgart; 1979 employed by Meister & Wittich, Stuttgart; since 1990 partnership with Norbert Diezinger; since 1999 professor at the FH Regensburg

Montessori College Oost in Amsterdam

Client: Stichting Montessori
Architect: Architectuurstudio Herman Hertzberger, Amsterdam
With: Willem van Winsen, Geert Mol, Arienne Matser, Henk de Weijer, Folkert Stropsma, Roos Eichhorn, Heleen Reedijk, Marijke Teijsse, Cor Kruter
Structural engineering:
Ingenieursgroep van Rossum
Date of construction: 1999

office@hertzberger.nl
www.hertzberger.nl

Herman Hertzberger
Born 1932; 1958 degree from the Technical University of Delft; 1959–69 editorial member of the Niederländischen Architekturzeitschrift Forum; 1965–69 teaching position at the Academy of Architecture in Amsterdam; 1970 associate professor at the Technical University of Delft; 1986 associate professor at the University of Geneva; 1990 Chairman of the Berlage Institute, Amsterdam; 1958 establishment of own office Amsterdam

Mini-golf and Ice Skating Hall in Bergheim

Client: Freibad Bergheim Errichtungs- und Betriebsgesellschaft m.b.H., Bergheim
Architects: mfgarchitekten, Graz
Friedrich Moßhammer,
Michael Grobbauer
Structural engineering:
JRCONSULT, Graz
Date of construction: 2005/2006

office@mfgarchitekten.at
www.mfgarchitekten.at

Friedrich Moßhammer
Born 1964 in Salzburg;
1994 degree from the TU Graz; 1995–2004 employed by Architects Riegler Riewe, Graz; since 2004 teaching position at the Institute for Architectural Technology TU Graz

Michael Grobbauer
Born 1966 in Anger, Austria; 1995 degree from the TU Graz; 1990–1996 employed by Wolff-Plottegg, Graz; 2003 Doktor der technischen Wissenschaften

Ice Stadium in Wolfsburg

Client: Stadtwerke Wolfsburg AG, Wolfsburg
Architects: Schulitz + Partner, Braunschweig
Helmut Schulitz, Marc Schulitz
With: Matthias Rätzel, Christian Laviola, Johannes König
Project management: Entricon GmbH, Wolfsburg
Structural engineering: Eilers + Vogel, Hanover; RFR Stuttgart (roof structure)
Date of construction: 2006

spa@schulitz.de
www.schulitz.de

Helmut C. Schulitz,
1962 degree from the TH Munich; 1974 establishment of the architectural office Schulitz + Partner in Los Angeles; since 1983 Schulitz + Partner in Braunschweig; 1969–1982 professor at the University of California; since 1982 professor at the TU Braunschweig

Marc Schulitz
Born 1974 in Santa Monica, USA; 1993–1999 studied architecture at various universtities; 1999 degree from the ETH Zurich; since 2000 employed by Schulitz + Partner, Braunschweig; since 2003 partner and since 2007 director of Schulitz + Partner, Braunschweig; since 2007 teaching position at the TU Braunschweig

Service Centre in Frankfurt am Main

Client: Ardi Goldman
Architects: Dietz Joppien Architekten AG, Frankfurt am Main
Project leader: Matthias Schönau
With: Thomas Kahmann, Torsten Herzog, Joachim Stephan, Nicole Weinbrecht, Sandra Große, Satura Wolff, Björn von Hayn
Structural engineering: Ingenieurbüro Phaidon Kostic, Frankfurt am Main
Date of construction: 2004

frankfurt@dietz-joppien.de
www.dietz-joppien.de

Albert Hans Dietz
Born 1958 in Saarbrücken; 1979–1986 studied architecture at the TH Darmstadt; 1986–1987 Master of Architecture at the University of Oregon; 2003–2004 teaching position at the Bergische University Wuppertal

Anett-Maud Joppien
Born 1959 in Frankfurt am Main; studied architecture at the TU Berlin and the TH Darmstadt; 1986 Master of Architecture at the University of California, Berkeley; since 2003 professor at the Bergische University Wuppertal

1997 establishment of the architectural office Joppien Dietz Architects in Frankfurt am Main; since 2004 Dietz Joppien Architects AG

Office and Training Centre in Dresden

Client: Bildungswerk der sächsischen Wirtschaft e.V., Dresden
Architects: Heinle Wischer, Dresden
Project leader: Jens Krauße
With: Michael Kraneis, Christian Hellmund, Nick Schreiter, Tobias Maschke
Structural engineering: Döking + Purtak Partnerschaft, Dresden
Date of construction: 2002

dresden@heinlewischerpartner.de
www.heinlewischerpartner.de

Thomas Heinle
Born 1961 in Stuttgart; 1986 degree from the Institute of Technology Biberach; since 1993 partner Heinle, Wischer und Partner, Freie Architekten GbR